Advances in Intelligent Systems and Computing

Volume 291

Series editor

Janusz Kacprzyk, Polish Academy of Sciences, Warsaw, Poland
e-mail: kacprzyk@ibspan.waw.pl

For further volumes:
http://www.springer.com/series/11156

About this Series

The series "Advances in Intelligent Systems and Computing" contains publications on theory, applications, and design methods of Intelligent Systems and Intelligent Computing. Virtually all disciplines such as engineering, natural sciences, computer and information science, ICT, economics, business, e-commerce, environment, healthcare, life science are covered. The list of topics spans all the areas of modern intelligent systems and computing.

The publications within "Advances in Intelligent Systems and Computing" are primarily textbooks and proceedings of important conferences, symposia and congresses. They cover significant recent developments in the field, both of a foundational and applicable character. An important characteristic feature of the series is the short publication time and world-wide distribution. This permits a rapid and broad dissemination of research results.

Carlos Ramos · Paulo Novais
Céline Ehrwein Nihan · Juan M. Corchado Rodríguez
Editors

Ambient Intelligence - Software and Applications

5th International Symposium on Ambient Intelligence

Springer

Editors
Carlos Ramos
Polytechnic of Porto
Porto
Portugal

Paulo Novais
CCTC/Informatics Department
University of Minho
Braga
Portugal

Céline Ehrwein Nihan
Haute Ecole d'Ingénierie et de Gestion
du Canton de Vaud
Yverdon-les-Bains
Switzerland

Juan M. Corchado Rodríguez
Dept. of Computing Science and Control
Faculty of Science
University of Salamanca
Salamanca
Spain

ISSN 2194-5357 ISSN 2194-5365 (electronic)
ISBN 978-3-319-07595-2 ISBN 978-3-319-07596-9 (eBook)
DOI 10.1007/978-3-319-07596-9
Springer Cham Heidelberg New York Dordrecht London

Library of Congress Control Number: 2014939995

Printed on acid-free paper

Springer is part of Springer Science+Business Media (www.springer.com)

Preface

This volume contains the proceedings of the 5th International Symposium on Ambient Intelligence (ISAmI 2014). The Symposium was held in Salamanca, Spain on June 4th-6th at the University of Salamanca, under the auspices of the Bioinformatic, Intelligent System and Educational Technology Research Group (http://bisite.usal.es/) of the University of Salamanca.

ISAmI has been running annually and aiming to bring together researchers from various disciplines that constitute the scientific field of Ambient Intelligence to present and discuss the latest results, new ideas, projects and lessons learned, namely in terms of software and applications, and aims to bring together researchers from various disciplines that are interested in all aspects of this area.

Ambient Intelligence is a recent paradigm emerging from Artificial Intelligence, where computers are used as proactive tools assisting people with their day-to-day activities, making everyone's life more comfortable.

After a careful review, 27 papers from 10 different countries were selected to be presented in ISAmI 2014 at the conference and published in the proceedings. Each paper has been reviewed by, at least, three different reviewers, from an international committee composed of 78 members from 24 countries.

Acknowledgments

Special thanks to the editors of the workshops: NTiAI .New Trends in Ambient Intelligence, CAIMaH. Challenges of Ambient Intelligence in the Workplace for the Management of Human Resources, AIfES .Special Session on Ambient Intelligence for Elderly Support.

We want to thank all the sponsors of ISAmI'14: IEEE Sección España, CNRS, AFIA, AEPIA, APPIA, AI*IA, and Junta de Castilla y León.

ISAmI would not have been possible without an active Program Committee. We would like to thanks all the members for their time and useful comments and recommendations.

We would like also to thank all the contributing authors and the Local Organizing Committee for their hard and highly valuable work.

Your work was essential to the success of ISAmI 2014.

June 2014

The Editors

Carlos Ramos
Paulo Novais
Céline Ehrwein Nihan
Juan M. Corchado Rodríguez

Organization

Organizing Committee Chair

Juan M. Corchado (Chairman)	University of Salamanca, Spain
Paulo Novais (Co-chairman)	University of Minho, Portugal

Program Committee

Carlos Ramos (Chairman)	Polytechnic Institute of Porto, Portugal
Ad van Berlo	Smart Homes, Netherlands
Andreas Riener	Johannes Kepler University Linz, Austria
Ângelo Costa	Universidade do Minho, Portugal
Antonio F. Gómez Skarmeta	University of Murcia, Spain
Antonio Fernández Caballero	University of Castilla-La Mancha, Spain
António Pereira	Instituto Politécnico de Leiria, Portugal
Benjamín Fonseca	Universidade de Trás-os-Montes e Alto Douro, Portugal
Carlos Bento	University of Coimbra, Portugal
Carlos Ramos	Polytechnic of Porto, Portugal
Catarina Reis	Instituto Politécnico de Leiria, Portugal
Cecilio Angulo	Polytechnic University of Catalonia, Spain
Cesar Analide	University of Minho, Portugal
Cristina Buiza	Ingema, Spain
Davy Preuveneers	Katholieke Universiteit Leuven, Belgium
Diane Cook	Washington State University, USA
Eduardo Dias	New University of Lisbon, Portugal
Elisabeth Eichhorn	Carmeq GmbH, Germany
Emilio S. Corchado	University of Burgos, Spain
Fernando Silva	Instituto Politécnico de Leiria, Portugal
Flavia Delicato	Universidade Federal do Rio de Janeiro, Brasil
Florentino Fdez-Riverola	University of Vigo, Spain
Francesco Potortì	ISTI-CNR institute, Italy

Francisco C. Pereira	SMART (Singapore-MIT Alliance) , Singapore
Francisco Silva	Maranhão Federal University, Brazil
Fulvio Corno	Politecnico di Torino, Italy
Goreti Marreiros	Polytechnic of Porto, Portugal
Gregor Broll	DOCOMO Euro-Labs, Germany
Guillaume Lopez	Aoyama Gakuin University, Japan
Habib Fardoum	University of Castilla-La Mancha, Spain
Hans W. Guesgen	Massey University, New Zealand
Ichiro Satoh	National Institute of Informatics Tokyo, Japan
Jaderick Pabico	University of the Philippines Los Baños, Philippines
Javier Carbo	Univ. Carlos III of Madrid, Spain
Javier Jaen	Polytechnic University of Valencia, Spain
Jo Vermeulen	Hasselt University, Belgium
Joel Rodrigues	University of Beira Interior, Portugal
José Carlos	Instituto Politécnico de Leiria, Portugal
José M. Molina	University Carlos III of Madrid, Spain
José Machado	University of Minho, Portugal
Joyca Lacroix	Philips Research, Netherlands
Juan A. Botía	University of Murcia, Spain
Juan R. Velasco	University of Alcala, Spain
Junzhong Gu	East China Normal University, China
Kasper Hallenborg	University of Southern Denmark, Denmark
Kristof Van Laerhoven	TU Darmstadt, Germany
Latif Ladid	University of Luxembourg, Luxembourg
Lawrence Wong Wai Choong	National University of Singapore, Singapore
Lourdes Borrajo	University of Vigo, Spain
Marco de Sa	Yahoo! Research, USA
Martijn Vastenburg	Delft University of Technology, Netherlands
Nuno Costa	Instituto Politécnico de Leiria, Portugal
Nuno Garcia	University of Beira Interior, Portugal
Óscar García	University of Salamanca, Spain
Pablo Haya	Universidad Autónoma de Madrid, Spain
Paul Lukowicz	University of Passau, Germany
Paulo Novais	University of Minho, Portugal
Radu-Daniel Vatavu	University "Stefan cel Mare" of Suceava, Romania
Rene Meier	Lucerne University of Applied Sciences, Switzerland
Ricardo Costa	Polytechnic of Porto, Portugal
Ricardo S. Alonso	University of Salamanca, Spain
Rui José	University of Minho, Portugal
Simon Egerton	Monash University, Malaysia
Teresa Romão	New University of Lisbon, Portugal

Thomas Hermann	Bielefeld University, Germany
Tibor Bosse Vrije	Universiteit Amsterdam, Netherlands
Veikko Ikonen	VTT Technical Research Centre, Finland
Vic Callahan	University of Essex, UK
Vicente Julián	Valencia University of Technology, Spain
Wolfgang Reitberger	Vienna University of Technology, Austria
Yi Fang	Purdue University, USA
Yoram Lev-Yehudi	Beacon Tech., Lda, Israel

Local Organization Committee

Dante I. Tapia	University of Salamanca, Spain
Javier Bajo	Polytechnic University of Madrid, Spain
Juan F. De Paz	University of Salamanca, Spain
Fernando De la Prieta	University of Salamanca, Spain
Gabriel Villarrubia González	University of Salamanca, Spain
Antonio Juan Sánchez Martín	University of Salamanca, Spain
Cesar Analide	University of Minho, Portugal
António Manuel de Jesus Pereira	Leiria Polytechnic Institute, Portugal

Workshops

NTiAI .New Trends in Ambient Intelligence

Organising Committee

Davide Carneiro	University of Minho, Portugal
Ângelo Costa	University of Minho, Portugal
José Carlos Castillo	University Carlos III, Spain

CAIMaH .Challenges of Ambient Intelligence in the Workplace for the Management of Human Resources

Organising Committee

Bajo, Javier	Technical University of Madrid
Boos, Daniel	Zürich
Borter, Silna	School of Business and Engineering Vaud
Corchado, Juan Manuel	University of Salamanca
Dufour, Florian	School of Business and Engineering Vaud
Ehrwein Nihan, Céline	School of Business and Engineering Vaud

Firoben, Laurence	School of Business and Engineering Vaud
Kinder-Kurlanda, Katharina E.	GESIS Cologne

AIfES .Special Session on Ambient Intelligence for Elderly Support

Organising Committee

António Pereira	Politechnic Intitute of Leiria, Portugal
Antonio Fernández-Caballero	Universidad de Castilla-La Mancha, Spain
Paulo Novais	University of Minho, Portugal

Contents

Context-Aware Automatic Service Selection Mechanism for Ambient Intelligent Environments

Houda Haiouni and Ramdane Maamri

Lire Laboratory, Constantine 2 University,
Constantine, Algeria
h.haiouni@gmail.com, rmaamri@yahoo.fr

Abstract. To develop context-aware Ambient Intelligence systems, suitable context models and reasoning approaches are necessary to provide the suitable services to users in dynamic and transparent manner. Due the advantages in modeling dynamic systems, Colored Petri Nets are adopted in this paper. So based on Colored Petri Nets modeling language, we propose a context-aware automatic service selection mechanism for intelligent environments. Using this formalism we also propose a solution to avoid conflict that can occur among resources sharing.

Keywords: Ambient Intelligence, Context Awareness, Multi Agent Systems, Colored Petri Nets, conflict resolution.

1 Introduction

The concept of Ambient Intelligence (AmI) has been introduced by European Commission's Information Society Technologies Advisory Group (ISTAG) [10]. AmI describes the future vision of computer science [23]. It deals with the vision that computing and communication ability are spread everywhere among nearly every object in our daily environment. An ambient system is a ubiquitous environment capable to interact intelligently with users and to provide the users with all the available functionalities and services in a *flexible*, *integrated* and almost *transparent* way for the end-user. For this purpose objects forming ambient systems must have some characteristics such as *autonomy*, *reactivity*, social abilities and *proactivity*.

The agent-based paradigm is one of the paradigms that can be used for the implementation of distributed systems. Typically, an agent has four properties [11]: *autonomy*, *social* ability, *reactivity* and *pro-activeness*. A multi-agent system (MAS) is a federation of agents interacting in a shared environment that cooperate and coordinate their actions given their own goals and plans. A MAS design can be beneficial in many domains, particularly when a system is composed of multiple entities that are distributed functionally or spatially [7]. Based on the above definitions, it is clear that agent paradigm is particularly appropriate for AmI. For these reasons, the use of MAS technology in ambient environments has been addressed in many researches like [2, 3], [5], [9], [21].

C. Ramos et al. (eds.), *Ambient Intelligence - Software and Applications*,
Advances in Intelligent Systems and Computing 291,
DOI: 10.1007/978-3-319-07596-9_1, © Springer International Publishing Switzerland 2014

To provide intelligent and pertinent services to users, AmI-based systems are expected to use contextual information such as location, identity, time, presence, temperature, etc. Dey and al [6] define the context as any information used to characterize the situation of an entity, which can be a person, a place or an object. A system is context-aware if it uses context to provide relevant information and/or services to the user, where relevancy depends on the user's task [8].

In this paper we focus both on context awareness and planning ability on ambient environments and we present a context-aware automatic service selection mechanism. Our proposed mechanism based on Colored Petri Nets (CPN) modeling language (an extension of Petri Nets (PN)). CPN are used to model contextual information and reasoning about. Based on this formalism we have proposed a solution to avoid conflict that can occur among resources sharing. The remaining part of this paper is organized as follows: section 2 shows a brief overview of the PN based context-awareness and MAS planning for AmI. The key concept of PN and CPNis given In section 3. In section 4, we present a context awareness modeling principals and we explain the reasoning process for context aware automatic services selection. An illustrative scenario is given in section 5. Finally, we recapitulate the main arguments and present some outline of future work.

2 Related Work

In this section, we briefly present some related work on multi-agent planning and Petri net-based context modeling for AmI systems.

Weld defines the planning problem as follows [24]. Given a description of the known part of the initial state of the world, a description of the goal (i.e., a set of goal stats), and a description of the possible actions that can be performed, modeled as state transformation functions. In the domain of AmI, to develop a planning mechanism we might take into consideration that distributed devices composing such systems are more compact in size, but their CPU and memory are much less powerful, and are often battery-powered. Consequently, these devices could not perform complex computations such those required by planning tasks. Because of AmI systems features, centralized planning is suitable. Many researches have proposed the use of man centralized components such [3, 4]. In [3] for example, the authors study what is the planner most appropriate for AmI systems. They have proposed the D-HTN (Distributed Hierarchical Task Network) planner. They based on the idea that, each smart spaces is equipped by an agent called majordomo that deals with the planning and execution activities and it is composed of one agent called device agent (split in the cooperative semiagent (CO) and the operative semiagent (OP)) for each device present in the environment. Really, for the required flexibility of the system and the number of devices involved, it is obvious that centralized control is not viable. However for limited smart spaces like smart homes, conferences rooms and many others, centralized planning could be adopted as solution. For systems with a large number of individuals we might push toward distributed planning. Authors in [19] present

Context-Aware Multi-Agent Planning (CAMAP), an approach for multi-agent planning that applies argumentation mechanisms to decide the most appropriate course of action according to the context information distributed among the agents. CAMAP is applied to AmI environment in the field of health-care.

Recently, the PN-based context-awareness modeling approaches were proposed in several works and they have been recognized as promising context models [20]. This is due mainly to both formal and graphical nature, expressiveness, and analytical property of PN. Modeling with PN inherently satisfies the requirements of context model, especially the usability of modeling formalisms and representation of relationships among context information [20]. Kwon [14] proposed an extension of CPN called Amended CPN to represent and analyze the context-aware systems. In this work the system is decomposed into several meaningful subsystems as a pattern. Amended CPN consist of multiple CPN, the Pattern and Context net represent contextual state and dynamic contextual change of another dimensional PN. As we proposed also in this work, in [14] each context type (location, weather, etc) is identified as color in Amended CPN and contextual information can be represented by a set of colors. Wang and Zeng [22] proposed a modeling methodology allows nondeterministic time duration for the activity. It permits to estimate the minimum and maximum duration time of each activity when the model is built. Moreover, it includes the resource constraints representing the resources which must be satisfied for executing the services. There are two kinds of places in the PN model for representing the activity and resources, Pa and Pr, respectively. A Pr containing a token means that the corresponding resource is available. Time and resource are critical ingredients in context-aware environment. This approach is one of the rare ones that consider both the time and resource constraints. However it presents few detailed information to represent resources constraints.

The related work presented above is summarized in Table 1. This paper contributes with the design of a model for Context-Aware Multi-Agent Planning, applied to Smart Home scenarios in AmI environments. Our approach uses CPN to model context and different services. We present also a solution to resource conflict occurring when services compete for the same resource in the same period of time.

Table 1. Summary of the related work

	AmI	*PN*	*MAS*	*Resources constraints*
[3]	Yes	No	Yes	No
[4]	Yes	No	Yes	No
[20]	Yes	No	Yes	No
[15]	Yes	Yes	Yes	No
[23]	Yes	Yes	No	Yes
Our approach	Yes	Yes	Yes	Yes

3 Concept of the Petri Nets

In contrast to (ordinary) PN, in which a token are uniform and represent typeless information, CPNs can carry complex information. They provide data typing (color sets) and sets of values of a specified type for each place. Formally a CPN [12] is a structure (P, T, A, N, ∑, V, C, E, G, M0), where:

- P, T, A are finite set of places, finite set of transitions and finite set of arcs, respectively. Such as $P \sqcap T = P \sqcap A = T \sqcap A = \emptyset$;
- N : is a node function such as : A→ (P x T ∪ T x P);
- ∑ : is a finite and non-empty set of data types, also called color sets or colors.
- V: is finite set of typed variables such as: TYPE [V] ∈ ∑;
- C: is a color set function. It assigns a color type (set) to each place: P→ ∑;
- E: A→ Expression E(a) of type C(p(a)), is the arc function. It maps each arc to an arc expression such: $\forall a \in A : [\text{Type}(E(a)) = C(p(a))_{MS} \wedge \text{Type}(\text{Var}(E(a))) \subseteq \Sigma]$ Where p(a) is the place of N(a);
- G: is a guard function. It assigns a guard to each transition. It is defined from T into Boolean expressions such that : $\forall t \in T : [\text{Type}(G(t)) = \text{boolean} \wedge \text{Type}(\text{Var}(G(t))) \subseteq \Sigma]$
- M0: is the initial marking. It is defined from P into expressions such that: $\forall p \in P : [\text{Type}(I(p)) = C(p)_{MS}]$

A transition t is said to be enabled if and only if the guard function G(t) attached to this transition is evaluated to true , and all of its input places are marked with at least E(p, t) tokens. The firing of an enabled transition t removes E(p, t) tokens from each input place p of t and adds E(t,p) tokens to each output place p of t.

4 Context-Aware Automatic Service Selection Mechanism

The multi agent architecture used for developing context awareness applications for smart spaces can contain mainly three types of agents: Planner Agent, Context Manager Agent and Sensor Agents (Fig 1). Agents communicate by sending and receiving messages following the hierarchy using communication language such as ACL (Agent Communication Language) language. *Sensor agents* residing in sensor level represent all perceptual components. Perceptual components are distributed through the environment, in order to provide a detailed model of the real world and make it available to the application. They provide elementary contextual perceptual information which is a key component of context awareness systems. In *Context Management Level*, an agent gathers data from different sensor agents. It aggregates sensor data over a time period. It can also deduce high-level context information from basic sensed contexts. The reasoning process is carried out by the Planner Agent in *Reasoning & decision level.* The reasoning about context is the process of mapping between context and services. Planner Agent uses information provided by Context manager Agent to select a set of possible services from Services library.

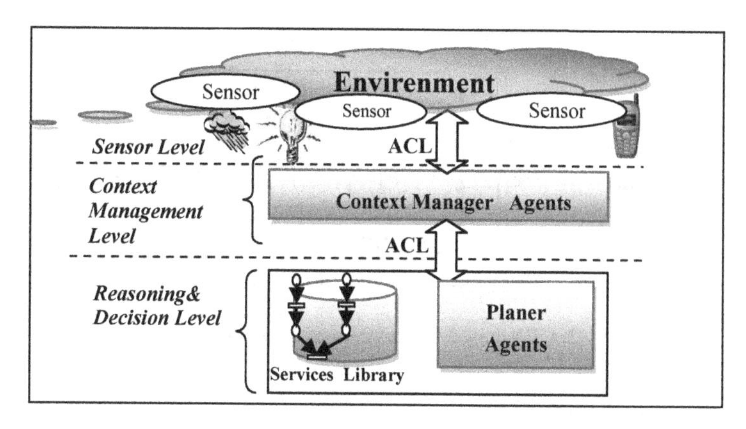

Fig. 1. Multi-Agent Architecture in Smart Home

4.1 Modeling Principles

The real world presents a dense multi-dimensional contextual space. Representing all of them is impossible for many reasons: i) this needs an enormous memory space; ii) some information is irrelevant or less important than others; ii) in practice, some type of context can't be captured, or it is expensive to capture them. So to develop context awareness applications, the first step consist of selecting form infinite contextual information list, those which are the interesting ones. In this setup, designers must determine which types of contextual information to be captured and to be used by the application.

After determining which information should be perceived and collected. The question now is how to represent context and how to reason about. Context modelling and reasoning are the core components of context-aware systems. In the following sub sections, we discuss of key concepts that can be modeled by CPN constructs (Fig. 2).

Context Modeling. Context-aware computing is the ability of applications to discover and react to changes in the environment in which they are situated (17). In order to represent the context and to reasoning about, we propose the adoption of CPN based model. With the flexibility of token definition it is possible to use them to model various contextual information. A context is a set of contextual information such as Context = C1* C2*...*Cn.

Resources Modeling. In the modeling of plans using CPN, we represent resources by resource places. We can use one resource place for each resource type as we can use one single resource place with a composite data type for convenience. In Fig. 1, PR1, PR2,..., PRn are resources places. An arc from action transition to resource place means that the execution of the action needs one (or more) resource (s) of that type.

Services Modeling. A service is a set of predefined ordering actions that refer to one (or more) contextual situation (s) (e.g. real world situation(s)). To model a service by CPN we use transitions to represent actions whereas places are used to represent actions stats. All service models will be stored in a plan library witch be used later by planner agent.

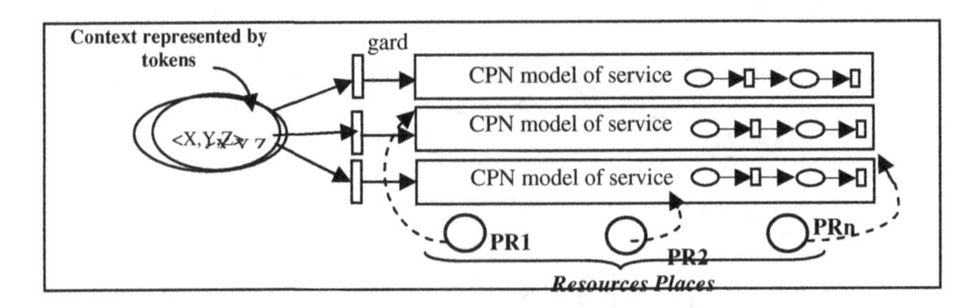

Fig. 2. CPN-based Context Awareness Modeling.

4.2 Reasoning about Context

Reasoning is used to decide what services should be invoked when any change of context occurs. Fig.3 shows the proposed service selection process:

Context Delivering & Management. In order to provide proactive and adequate services to the user, context delivering components have to be integrated. The role of these components is to continuously track information about user and his environment. Context management includes many operations such as sensor data aggregation over a time period and inferring high-level context information from basic sensed contexts …etc.

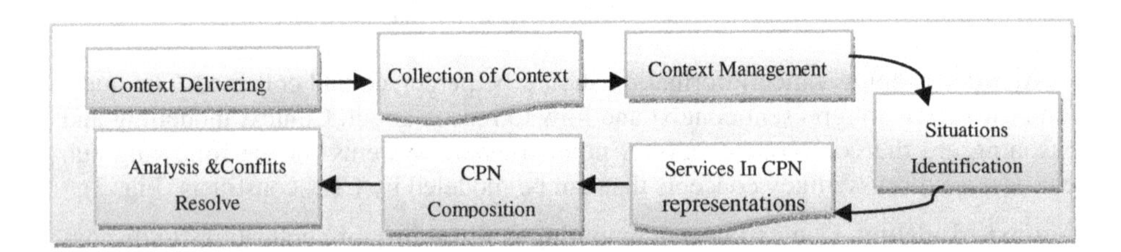

Fig. 3. Services Selection Process

Situations Identification. The reasoning about context is the process of mapping between context and services. A situation can be defined by collecting relevant contexts, uncovering meaningful correlations between them, and labeling them with a descriptive name. The descriptive name can be called a descriptive definition of a situation, which is about how a human being defines a state of affairs in reality [13]. Situation Identification is the processes to deriving a situation by interpreting or merging several pieces of perceived information. Situation identification techniques have been studied extensively in AmI. The most promise one have been highlighted in [13]. In this stage all possible contextual situations will be identified. Each contextual situation represents a service modeled by CPN. Services models are created by designers of context-aware applications. According to token values, all possible services models will be selected from service models library. The Planner Agent will have *n* CPN service models (*n* is positive integer).

CPN Composition & Conflict Resolution. It is possible that for some context values many services can be invocated. However in certain cases conflict among resources sharing can bloc system evolution. In the other hand, in many cases some types/values of information are more important than others, or in other words not all services have the same importance. As an example, consider the tow following information: "the gas range is turned on for 10units of times" and "weather is raining". In this situation, conflict among resources which are windows could happen. Because it is raining windows must be closed. In the other hand the windows must be opened because the gas range has been turned on (if there are not other kinds of aeration). From our point of view, this phenomenon can be avoided by *Establishing Priority for Services*. To do that, we propose in this framework to classify services into several types. For each service class we associate a priority level as fallow: *class 1 → level 0 (higher priority), class 2 → level 1 ,..., and class n → level n-1 (lowest priority)*

The priority mechanism can be applied using CPNby assigning priority to transitions. We choose to define priority as a positive real-valued function over transitions, the higher the value, the greater the priority. So we define the priority function ρ mapping a transition into R^+. When two transitions are enabled, the one with the highest priority fires. All transitions forming the same service model have the same priority. The role of priority transitions appears in merging services models phase where conflicts among resources can take place. After applying priorities and merging CPN Models the resultant net will be executed to obtain the order in which a set of possible services are executed.

5 Illustrative Scenario: Smart Home

AmI can be applied in any dynamic environment where is a need to manage tasks and automate services (e.g., hospitals schools, homes, etc). Abascal et al [1] assumes that home is the ideal place to apply percepts and technologies for giving high- level services to the user. Smart homes represent the ideal solution for individuals with different needs and abilities (child, old, blind, handicap, etc). So we illustrate our approach through smart home scenario. The main expected benefits of this technology can be: i) *Increasing safety*: e.g., by automating specific tasks that an individual with disabilities or elderly can't perform them. Or by providing a safe and secure environment;ii)*Comfort*: e.g., by adjusting light, temperature , TV channels automatically); iii)Economy: e.g., by monitoring the use of energy. To ensure these benefits, the system must provide many services. Services can be classified into three classes according to their objective. The first class represents services providing safety or security. The second class represents all services that provide users comfort, and the last one represents services having economic aspect. From our point of view, assuring safety of peoples has much important than assuring there comfort. For this reason, the first service class must have the higher priority. Concerning the second and the third ones, we assume that comfort aims became before economy one. For instance, if the user prefers high lighting, and energy module detects an over use of energy, the systems will give more importance to user's preferences. The house is equipped with smart devices and sensors forming an AmI system. Imagine now that the owner of the house is sleeping. The weather is very could in outside (-5°), all windows all closed, and carbon monoxide (CO) alarms detects poisonous CO gas in home. The system uses this perceived information to select a set of possible services from Services library. In this case, tow services can be invoked. Fig. 4, Fig. 5 represent CPN model of selected services (designed with CPN tools [25]).

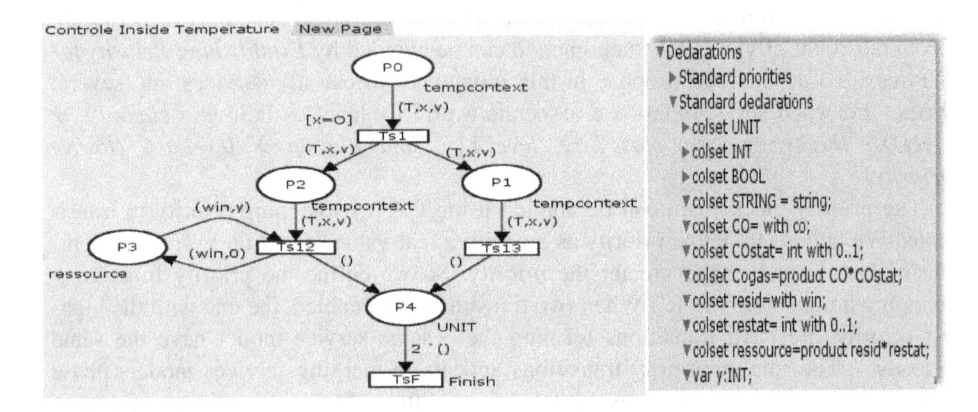

Fig. 4. Control Inside Temperature

The first service aims to regulate inside temperature. It is composed of the tow flowing actions: i) Close windows, ii) regulate inside temperature by increasing/or decreasing it. The second one aims to protect inhabitant life by performing flowing actions: i) windows must be opened; ii) occupant must be informed as soon as possible.

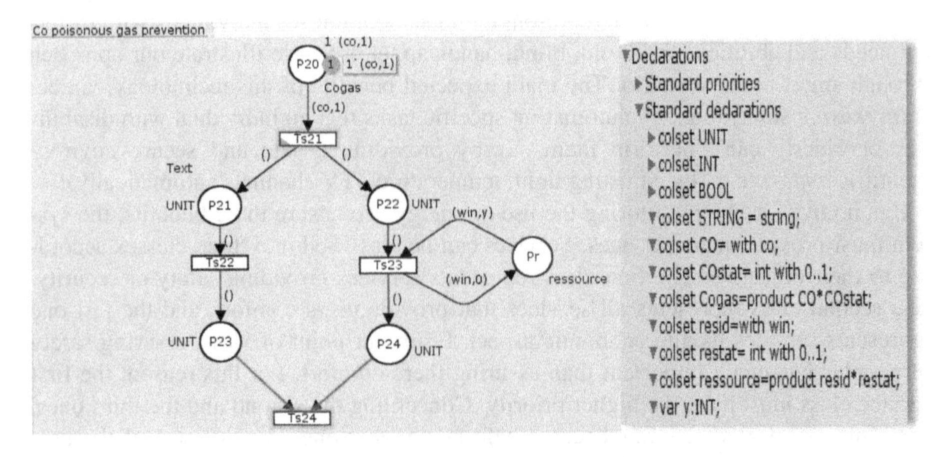

Fig. 5. CO Poisonous Gas Prevention

The CPN model presented in Fig. 6 is obtained by merging service 1 and service2 CPN models. Note that both services compete for the same resource (*Pr* place) simultaneously. Both of them pretend to take the control over windows. Our solution to this conflict pass through using priorities associating to services to choose which one has the privilege of accessing first the resource. Because service 2 deals with unsafe situation it will have the higher priority. So all transitions of service 2 model (Ts21, Ts22, Ts23) will have the higher priority, $\rho\ (T)=1$, and for all transitions *t'* of service 2 CPN model, $\rho\ (T)=0$. As consequence service 2 will be executed before service 1.

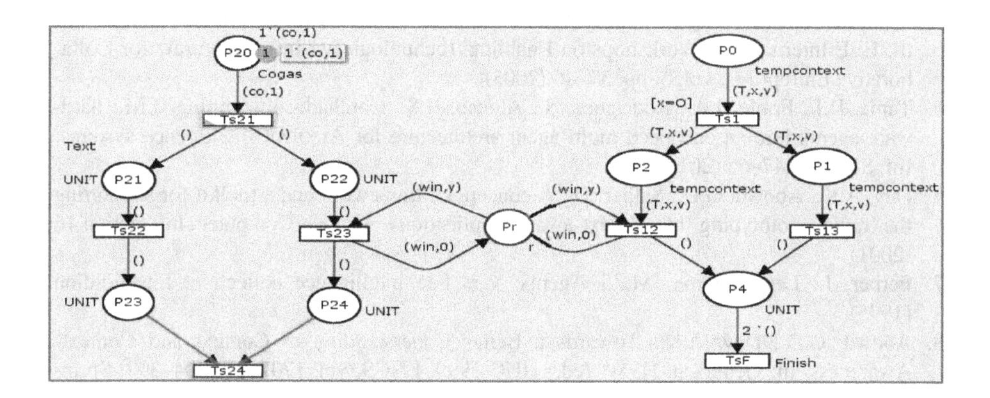

Fig. 6. The resultant Merged CPN

6 Conclusion and Future Work

In this paper we have proposed context-aware automatic service selection approach in intelligent environment. PN have been successfully used for modeling and analysis behavior of various complex dynamic systems. For this reason we have adopted CPN as modeling language to model contextual information and reasoning about. We have proposed a solution to avoid conflict that can occur among resources sharing. Our solution based on the idea of establishing priorities to services. Using CPN formalism, this idea has been represented by associating priorities to transitions. Future research would investigate the establishment of priorities in a flexible manner rather than being done during the development process. In this work only the conflict occurring among resources sharing has been taken into account. However, Conflicts can occur for many others reasons in AmI systems [25]. For instance, when there are conflicting user preferences in the same moment, e.g. One user prefers a lamp to be turned on while another user prefers that it should be turned off. So our efforts now are concentrated on expending the proposed reasoning process based on a formal model to detect other kinds of conflicts on AmI environments and resolve them.

References

1. Abascal, J., Civit, A., Fellbaum, K.: Smart Homes- Technologiy for the future. In: COST Seminar, Leuven, Belguim. Telecommunications: Access for all (2001)
2. Abraham, A., Corchado, J.M., Alonso, R.S.: Agents and ambient intelligence: case studies. In: Dante, I., Tapia, J. (eds.) Ambient Intell. Human Comput., vol. 1, pp. 85–93 (2010), doi:10.1007/s12652-009-0006-2
3. Amigoni, F., Gatti, N., Pinciroli, C., Roveri, M.: What planner for ambient intelligence applications? IEEE Transactions on Systems, Man and Cybernetics, Part A: Systems and Humans 35(1), 7–21 (2005)
4. Carbi, G., Ferrari, L., Leonardi, L., Zambonelli, F.: The LAICA project/ supporting ambient intelligence via agents and ad-hoc middelware. In: Proceedings of WETICE 2005, 14 th IEEE International Workshops on Enabling Technologiesw/ Infrastructurew for Collaborative Entreprises, vol. 5, pp. 39–46 (2005)
5. Tapia, D.I., Fraile, J.A., Rodrı́guez, S., Alonso, R.S., Corchado: Integrating, J.M.: hardware agents into an enhanced multi-agent architecture for Ambient Intelligence systems. Inf. Sci. 222, 47–65 (2013)
6. Dey, A.K., Abowd, G.D., Salber, D.: A conceptual framework and a toolkit for supporting the rapid prototyping of context-aware applications. Human-Computer Interaction 16 (2001)
7. Ferber, J.: Les Systèmes Multi Agents: vers une intelligence collective. InterEdidion (1995)
8. Abowd, G.D., Dey, A.K.: Towards a Better Understanding of Context and Context-Awareness. In: Gellersen, H.-W. (ed.) HUC 1999. LNCS, vol. 1707, pp. 304–307. Springer, Heidelberg (1999)
9. Hu, X., Du Bruce Spencer, W.: A Multi-Agent Framework for Ambient Systems Development. In: MobiWIS 2011 (2011)
10. ISTAG, Scenarios for Ambient Intelligence in 2010, European Commission Report (2001)
11. Jennings, N.R., Wooldridge, M.J.: Applications of intelligent agents. In: Nicholas, R. (ed.)
12. Jensen, K.: Coloured Petri nets and the invariant method. T.S.C. 14, 317–336 (1981)
13. Ye, J., Dobsona, S., McKeever, S.: Situation identification techniques in pervasive computing: A review. Pervasive and Mobile Computing 8, 36–66 (2012)
14. Kwon, O.B.: Modeling and generating context-aware agent-based applications with amended colored petri nets. Exp. Syst. Appl. 27(4), 609–621 (2004)
15. Murata, T.: Petri nets: properties, analysis and applications. Proceedings of the IEEE 6(1), 39–50 (1990)
16. Peterson, J.K.: Petri Nets Theory and the modeling of systems. Prentice-hall.inc., Englewood ellifs (1981)
17. Satoh, I.: Mobile Agents for Ambient Intelligence. In: Ishida, T., Gasser, L., Nakashima, H. (eds.) MMAS 2005. LNCS (LNAI), vol. 3446, pp. 187–201. Springer, Heidelberg (2005)
18. Schilit, B.N., Theimer, M.M.: Disseminating active map information to mobile hosts. Netw. IEEE 8(8), 22–32 (1994)
19. Ferrando, S.P., Onaindia, E.: Context-Aware Multi-Agent Planning in intelligent environments. Information Sciences 227, 22–42 (2013)
20. Han, S., Youn, H.Y.: Petri net-based context modeling for context-aware systems. Springer Science+Business Media B.V (2011)

21. Vallée, M., Ramparany, F., Vercouter, L.: Composition Flexible de Services d'Objets Communicants. In: UbiMob 2005: Proceedings of the 2nd French-Speaking Conference on Mobility and Ubiquity Computing, pp. 185–192 (2005)
22. Wang, H., Zeng, Q.: Modeling and analysis for workflow constrained by resources and nondetermined time: an approach based on petri nets. IEEE Trans. Syst. Man Cybern. A Syst. Hum. 38(4), 802–817 (2008)
23. Weiser, M.: The computer for the 21st century. Scientific American 256(3), 94–104 (1991)
24. Weld, D.S.: Recent advances in AI planning. AI Magazine 20(2), 93–123 (1999)
25. Carreira, P., Resendes, S., Santos, A.: Toward automatic conflict detection in home and building automation systems. Pervasive and Mobile Computing (2013)
26. CPN Tools, http://cpntools.org/

A Multimodal Conversational Agent for Personalized Language Learning

David Griol, Ismael Baena, José Manuel Molina, and Araceli Sanchis de Miguel

Computer Science Department
Carlos III University of Madrid
Avda de la Universidad, 30, 28911 - Leganés, Spain
{david.griol,josemanuel.molina,araceli.sanchis}@uc3m.es,
100065131@alumnos.uc3m.es

Abstract. Conversational agents have became a strong alternative to enhance educational systems with intelligent communicative capabilities. In this paper, we describe a multimodal conversational agent that facilitates an independent and user-adapted second language learning. The different modules of the system cooperate to interact with students using spoken natural language and visual modalities, and adapt their functionalities taking into account their evolution and specific preferences. The results of a preliminary evaluation show that users' satisfaction with the system was high, as well as the perceived didactic potential and adaptive functionalities.

Keywords: Multimodal conversational agents, e-learning, educative technology, natural language processing.

1 Introduction

Ambient Intelligence is characterized by intelligent, pervasive, and seamless computer systems embedded into everyday devices, tailored to the individual's context-aware needs and providing a natural and intelligent interaction. This way, multimodal conversational agents [1] have became a strong alternative to enhance multi-agent systems with these intelligent communicative capabilities [2].

With the growing maturity of conversational technologies, the possibilities for integrating conversation and discourse in e-learning are receiving greater attention. Using natural language in educational software allows students to spend their cognitive resources on the learning task, and also develop more social-based agents [3].

Current possibilities to employ conversational agents for educative purposes include tutoring applications [4], question-answering [5], conversation practice for language learners [6], pedagogical agents and learning companions [7], and dialogs to promote reflection and metacognitive skills [8]. These agents may also be used as role-playing actors in immersive learning environments [9].

C. Ramos et al. (eds.), *Ambient Intelligence - Software and Applications*,
Advances in Intelligent Systems and Computing 291,
DOI: 10.1007/978-3-319-07596-9_2,

Systems developed to provide these functionalities typically rely on a variety of components, such as speech recognition and synthesis engines, natural language processing components, dialog management, databases management, and graphical user interfaces. Laboratory systems usually include specific modules of the research teams that build them, which make portability difficult. Thus, it is a challenge to package up these components so that they can be easily installed by novice users with limited engineering resources. In addition, due to this variability and the huge amount of factors that must be taken into account, these systems are difficult to develop and typically are developed ad-hoc, which usually implies a lack from scalability. Our work represents a step in this direction.

In this paper we describe a multimodal conversational agent for adaptive second language learning. The system has been developed by means of a modular approach that allows to easily developing multimodal conversational agents for pedagogical applications. This approach facilitates a rapid and cost-effective development. This way, different alternatives can be considered for each module, and the pedagogic knowledge is separated from the technical details, so that teachers and parents can add new contents without having a technical background at the same time as the software includes these new data for the interaction with the students.

2 The *Test Your English* Pedagogical System

The *Test Your English* pedagogical system has been designed with the main aim of facilitating an independent and personalized second language learning. Figure 1 shows the initial screen of the system. As it can be observed, users must register in the application. This way, their previous interactions can be taken into account to provide user adaptation functionalities.

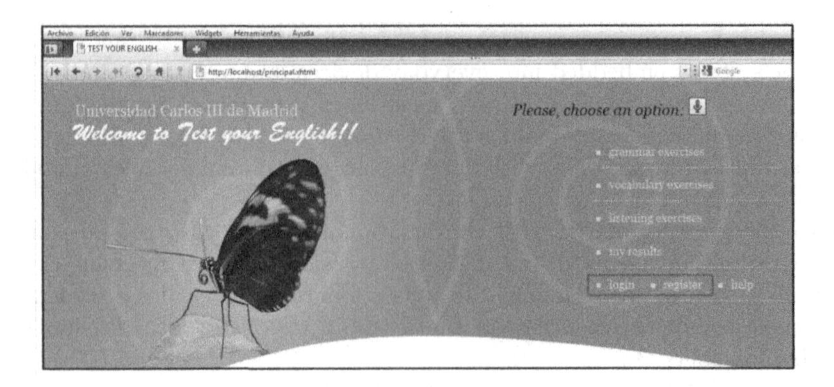

Fig. 1. Main screen of the *Test Your English* system

The current version of the system is web-based. The application consists of the set of core components for multimodal dialog systems (speech recognition, spoken language understanding, dialog management, language generation, and speech synthesis) and split across a server and a client device running a standard web browser. Core components on the server provide speech recognition and speech synthesis capabilities, access to the databases, and a logger component which records user interactions.

The system is accessible to any user with a standard web browser and a network connection. This way, the application can be easily accessed not only from desktop, laptop, and tablet computers, but from a variety of mobile devices as well. In addition, an Audio Controller component runs on the client to capture a user's speech and stream it to the speech recognizer, as well as to play synthesized speech generated on the server and streamed to the client.

Natural language understanding is performed by means of grammars which include the different options that are also visually provided to the student. To do this, we follow the Java Speech Grammar Format (JSGF, `www.w3.org/TR/jsgf/`), which allows specifying these sentences in a compact way, easily adapted and also embedding semantics into the grammar.

Speech recognition hypotheses are passed to the dialog manager architecture for processing. Regarding dialog management, all the events in the application are controlled using JavaScript. Given the requisites of the task, we decided to use a dialog model based on finite states in which at each moment a question is selected and shown in the screen along with the alternative answers and the associated multimedia files.

The main functions of the system can be classified into three main modules: Practice, Assessment, and Contents management. The Practice module includes three kinds of exercises: grammar, vocabulary, and listening exercises. Students can access this module by means of a form in which they can select the level and category of the exercises (e.g., verbal tenses for the grammar exercises, topic for the vocabulary exercises, or title of the text for the listening exercises).

Grammar exercises (see Figure 2 top-left) consist of filling gaps in sentences with a verb, adjective, adverb or other grammar elements from the topic and difficulty level initially selected. Vocabulary exercises (Figure 2 top-right) allow users to dictate or write words that are described by means of images displayed to the user. Listening exercises consist of two main parts (Figure 2 bottom). Firstly, the multimodal system reads the text selected by the user. Then, a set of related questions are presented to the user to evaluate their reading comprehension.

The Assessment module (see Figure 3) allows users to review the answers that they provided to the previously selected exercises, visualize the correct solution in case of errors, and know detailed statistics about the student's specific evolution using the system.

The personalization of the system is carried out by means of the "choose for me" functionality of the Practice module. This functionality takes into account the number of exercises in each category and level correctly solved by each user, so that the system can provide personalized suggestions and select the following

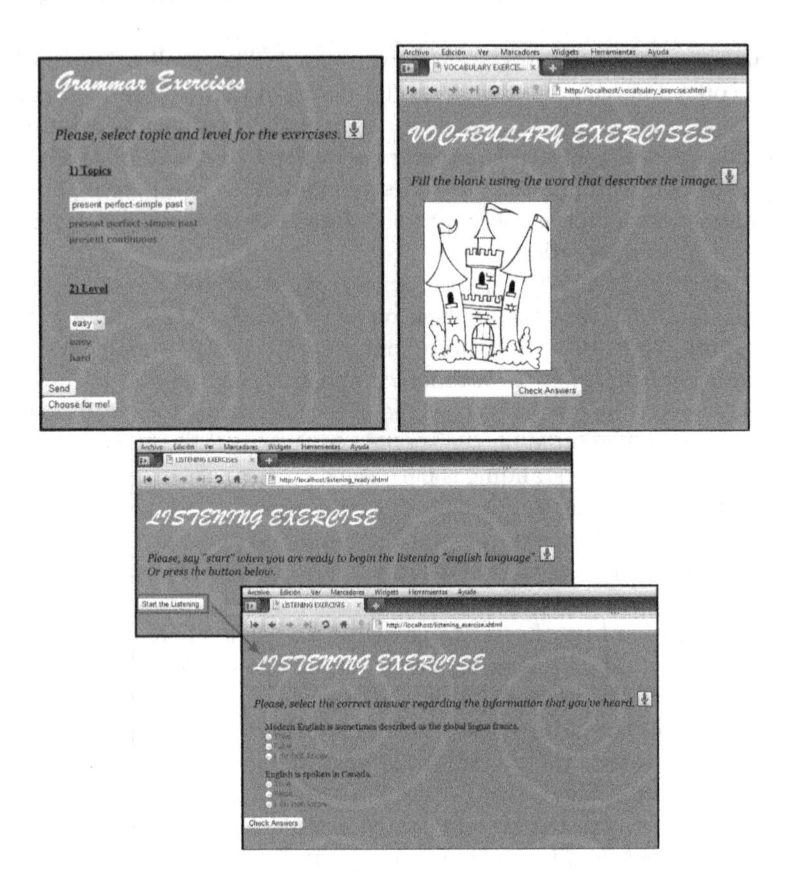

Fig. 2. Example of the different kinds of exercises offered by the system

exercises according to the errors found in previous interactions. To do this, the system manages the database containing the exercises completed by the user for the current category and difficulty level. The system also analyzes whether the number of mistakes made for the current difficulty level is highest than a specific threshold (initially predefined to the half of them). The objective is to personalize users' recommendations taking into their specific evolution with the different categories of exercises and difficulty levels.

Figure 4 shows a set of use cases for the personalization functionality of the system. In the first case, the system recommends trying additional exercises from the same difficulty level that the previously selected given that the student has not already successfully completed at least the half of them. In the second case, the system provides the same recommendation given that the number of mistakes is higher than a specific threshold. In the third case, the system suggests selecting the next difficulty level given that the results for the current level are satisfactory.

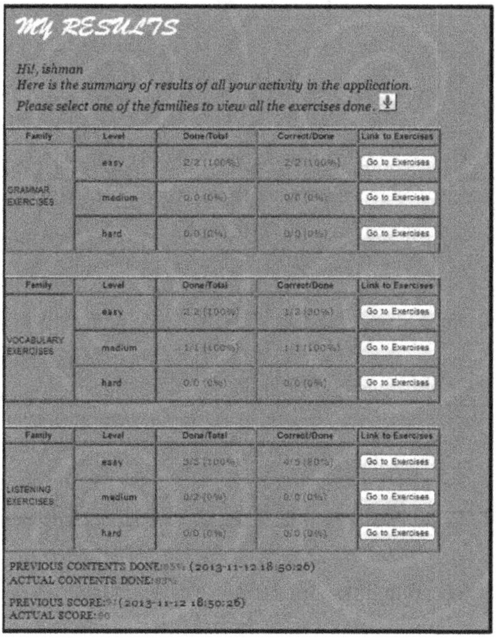

Fig. 3. Assessment module of the *Test Your English* system

Finally, the Contents management module allows administrator users to modify and insert new contents in the system. This module is based on different functionalities provided by the phpMyAdmin tool (`www.phpmyadmin.net`) to facilitate creating new exercises and editing or removing existing ones. The system comprises three main databases that contain the learning contents, multimodal elements and the history of the interaction respectively. The first database stores the questions and answers categorized in different topics. For each question, there is a text, optional multimedia contents (audio and video) and several answers. For each answer, there is also text and/or multimedia, as well as the positive and negative feedbacks and hints to be provided to the student in the case he/she selects the answer. For each question only one answer is assumed to be correct. The second database contains the visual rendering of the interface, and the third database stores the information about the previous interactions of the user with the system.

The objective is to facilitate including new questions and editing the existing ones. This way, different people can help in the development of the system without requiring an expert knowledge in conversational agents. For example, teachers and parents can include new questions in the database, and graphic designers can create attractive multimodal contents for the system and include them into the corresponding databases.

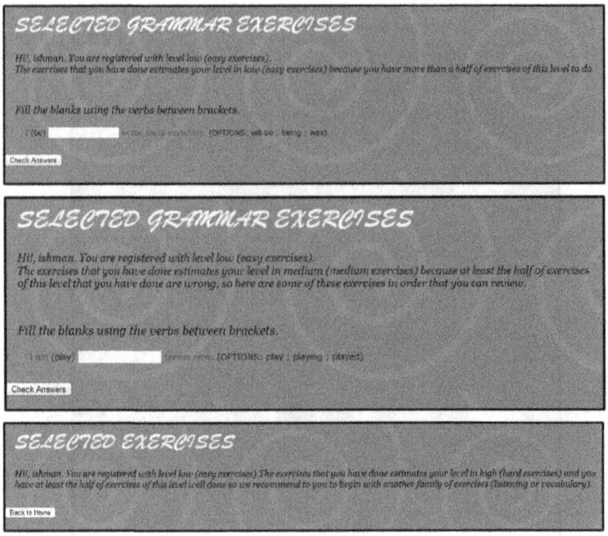

Fig. 4. Use cases describing the operation of the *choose for me* functionality

3 Preliminary Evaluation

We have already completed a preliminary evaluation of the developed pedagogic system. A total of recruited 25 users participated in the evaluation, aged 21 to 57 (mean 27.2), 17 male and 8 female. Prior to the evaluation, an assistant explained to the users what the system is about and how it is used. Then, they were given 30 minutes to get accustomed to the system, while the doubts were solved by a teacher and the assistant. To do this, each user had at his/her disposal a computer and had to wear a microphone headset. They were allowed to break up the interaction at any time for any reason. Then, each user freely interacted with the application during 10 sessions of at list 15 minutes.

At the end of the last session each user was required to complete the questionnaire shown in Table 1 was defined for the evaluation. The responses to the questionnaire were measured on a five-point Likert scale ranging from 1 (strongly disagree) to 5 (strongly agree). The users were also asked to rate the system from 0 (minimum) to 10 (maximum) and there was an additional open question to write comments or remarks.

Also, from the interactions of the experts with the system we completed an objective evaluation of the application considering the following interaction parameters: i) Question success rate (SR). This is the percentage of successfully completed questions: system asks - user answers - system provides appropriate feedback about the answer; ii) Confirmation rate (CR). It was computed as the ratio between the number of explicit confirmations turns and the total of turns; iii) Error correction rate (ECR). The percentage of corrected errors.

Table 1. Questionnaire employed for the subjective assessment

Interaction experience and technical quality
IT01. The system is easy to use
IT02. The system provides adequate feedback
IT03. The system is helpful
IT04. The system offers enough interactivity
IT05. It is easy to know what to do at each moment
IT06. The amount of information that is displayed on the screen is adequate
IT07. The system is adapted to my learning degree
IT08. I would use the system again
Learning contents and didactic potential
LD01. The questions were easy to understand
LD02. The questions were easy to answer
LD03. The system help me to learn new things
LD04. The activities support significant learning
LD05. The feedback provided by the agent improves learning
LD06. The system encourages continuing learning after errors
LD07. The system made me appreciate my skills for learning English

Table 2. Results of the evaluation of the system

	Min / max	Average	Std. deviation
IT01	3/4	3.87	0.26
IT02	4/5	4.73	0.40
IT03	4/5	4.85	0.34
IT04	3/5	4.06	0.73
IT05	4/5	4.86	0.31
IT06	4/5	4.93	0.15
IT07	3/5	4.52	0.41
IT08	5/5	5.00	0.00
LD01	3/5	4.26	0.55
LD02	3/5	4.33	0.39
LD03	4/5	4.85	0.22
LD04	5/5	5.00	0.00
LD05	4/5	4.77	0.59
LD06	4/5	4.85	0.38
LD07	3/5	4.32	0.41

SR	CR	ECR
93.05%	17.25%	91.92%

The results of the questionnaire are summarized in Table 2. As can be observed, the system was rated fairly well and most of the users learned new contents and most of them would like to use again the system. The satisfaction with technical aspects was high, as well as the perceived didactic potential. The system was considered attractive and adequate and the users felt that the system is appropriate and the activities relevant. The global rate for the system was 8.7 (in the scale from 0 to 10).

Although the results were very positive, in the open question the users also pointed out desirable improvements. One of them was to make the system listen constantly instead of using the push-to-talk interface. In fact, an analysis of the main problems detected showed that most of these errors were due to the users not holding the push-to-talk button correctly and thus the input was cut, or because they used longer phrases or fillers which were not correctly processed by the system. However, in most cases, these problems could be overcome by confirming or asking again for the data, as shown by the question success rate of 93.05%. Additionally, the approaches for error correction by means of confirming or re-asking for data were successful in 91.92% of the times when the speech recognizer did not provide the correct answer.

4 Conclusions

According to previous works, multimodal conversational agents can accelerate the learning process, facilitate access to education, personalize the learning process, and supply a richer learning environment. These important points are usually addressed by establishing a more engaging and adaptive relationship between the students and the system. In this paper, we have described the *Test Your English* multimodal conversational agent, which has been developed to provide this enhanced educative environment for second language learning. The system is comprised of different modules that cooperate to interact with students using speech and visual modalities, and adapt its functionalities taking into account their evolution and specific preferences.

Although there are currently many systems for students to learn a second language, most of them are designed to follow the same behavior for every student, not taking into account their specific evolution during the learning process. The experimental results show that the adaptation provided by our system and the natural communication that it provides have a very positive impact on the learning outcomes and satisfaction of the students. For future work, we plan to replicate the experiments with more students to validate these preliminary results and incorporate their suggestions.

Acknowledgements. This work was supported in part by Projects MINECO TEC2012-37832-C02-01, CICYT TEC2011-28626-C02-02, CAM CONTEXTS (S2009/TIC-1485).

References

1. Pieraccini, R.: The Voice in the Machine: Building Computers that Understand Speech. The MIT Press (2012)
2. Corchado, J., Tapia, D., Bajo, J.: A multi-agent architecture for distributed services and applications. Computational Intelligence 24(2), 77–107 (2008)
3. Rodríguez, S., de Paz, Y., Bajo, J., Corchado, J.M.: Social-based Planning Model for Multi-agent Systems. Expert Systems with Applications 38(10), 13005–13023 (2011)
4. Pon-Barry, H., Schultz, K., Bratt, E.O., Clark, B., Peters, S.: Responding to student uncertainty in spoken tutorial dialog systems. Int. Journal of Artificial Intelligence in Education 16, 171–194 (2006)
5. Wang, Y., Wang, W., Huang, C.: Enhanced Semantic Question Answering System for e-Learning Environment. In: Proc. AINAW, pp. 1023–1028 (2007)
6. Fryer, L., Carpenter, R.: Bots as Language Learning Tools. Language Learning and Technology 10(3), 8–14 (2006)

7. Cavazza, M., de la Camara, R.S., Turunen, M.: How Was Your Day? a Companion ECA. In: Proc. AAMAS 2010, pp. 1629–1630 (2010)
8. Kerly, A., Ellis, R., Bull, S.: CALMsystem: A Dialog system for Learner Modelling. Knowledge Based Systems 21(3), 238–246 (2008)
9. Griol, D., Molina, J., Sanchis, A., Callejas, Z.: A Proposal to Create Learning Environments in Virtual Worlds Integrating Advanced Educative Resources. Journal of Universal Computer Science 18(18), 2516–2541 (2012)

Power Indices of Influence Games and New Centrality Measures for Agent Societies and Social Networks*

Xavier Molinero[1], Fabián Riquelme[2], and Maria Serna[2]

[1] Dept. of Applied Mathematics III. UPC, Manresa, Spain
xavier.molinero@upc.edu
[2] Dept. de Llenguatges i Sistemes Informàtics. UPC, Barcelona, Spain
{farisori,mjserna}@lsi.upc.edu

Abstract. We propose as centrality measures for social networks two classical power indices, Banzhaf and Shapley-Shubik, and two new measures, *effort* and *satisfaction*, related to the spread of influence process that emerge from the subjacent influence game. We perform a comparison of these measures with three well known centrality measures, *degree*, *closeness* and *betweenness*, applied to three simple social networks.

Keywords: Social Network, Centrality, Power index, Influence game, Simple game.

1 Introduction and Preliminaries

We propose to study networked societies, social networks or agent societies, from the social networking point of view. Social network analysis is a multidisciplinary field related to sociology, computer science and mathematics, among other topics. One of the most studied concepts is *centrality*, that measures how structurally important is an actor within a social network [4,15,8,12]. Here we consider seven centrality measures: Banzhaf and Shapley-Shubik power indices through the use of *influence games* [10]; two new measures, the *effort* and the *satisfaction*; and the classic ones [8], *degree*, *closeness* and *betweenness*. We perform an experimental comparison on three simple real social networks, *monkeys' interaction* [3,8], *dining-table partners* [11,2], and *student Government discussion* [6,2].

A *social network* is a directed edge-labeled graph (G, w), where $G = (V, E)$ is a graph without loops, V is the set of nodes representing individuals, actors, players, etc., E is the set of edges representing interpersonal ties between actors, and $w : E \to \mathbb{R}$ is a *weight function* which assigns a weight to every edge, representing the strength of each interpersonal tie. An actor $i \in V$ has *influence* over another $j \in V$ if and only if $(i, j) \in E$.

* This work was partially supported by 2009SGR1137 (ALBCOM). X. Molinero, F. Riquelme and M. Serna are also partially funded by grants MTM 2012-34426, BecasChile (CONICYT), and TIN2007–66523 (FORMALISM), respectively.

C. Ramos et al. (eds.), *Ambient Intelligence - Software and Applications*,
Advances in Intelligent Systems and Computing 291,
DOI: 10.1007/978-3-319-07596-9_3,

Now we consider three of the most well-known (normalized) centrality measures [8], which study the relevance of a node inside a network [15]. We use the notation $deg^-(i) = |\{j \in V \mid (j,i) \in E\}|$ and $deg^+(i) = |\{j \in V \mid (i,j) \in E\}|$.

Degree centrality (C_D): measures the average indegree or outdegree of each actor, $C_D^-(i) = deg^-(i)/(n-1)$, or $C_D^+(i) = deg^+(i)/(n-1)$. For undirected networks, $deg(i) = deg^-(i) = deg^+(i)$, so we set $C_D = C_D^- = C_D^+$.

Closeness centrality (C_C): It is based on the inverse of the sum of the shortest distances from i to the other actors. Let D be the usual distance matrix of the network in which, if there is no path from i to j, we set $(D)_{ij} = n$. We define $C_C(i) = (n-1)/\sum_{i \neq j}(D)_{ij}$.

Betweenness centrality (C_B): Let b_{jk} the number of shortest paths from the node j until k, and b_{jik} the number of these shortest paths that pass through i. If there is no path from j to k, we assume that $b_{jik}/b_{jk} = 0$. We define $C_B(i) = \sum_{j \neq k} \frac{b_{jik}}{b_{jk}}/((n-1)(n-2))$.

Notation related to simple and influence games comes from [13,10]. An *influence graph* is a tuple (G, w, f) where (G, w) is asocial network and $f : V \to \mathbb{N}$ a labeling function that quantifies how influenciable each actor is.

Given an influence graph (G, w, f) and an initial activation set $X \subseteq V$, the *spread of influence* [7], in the linear threshold model, is denoted by $F(X)$, where $F(X) \subseteq V$ is formed by the actors activated through an iterative process in which initially only the nodes in X are activated. Let $F^t(X)$ be the set of nodes activated at some iteration t, then at the next $t+1$ iteration a node $i \in V$ will be activated iff $\sum_{j \in F^t(X)} w((j,i)) \geq f(i)$. The process stops when no additional activation occurs.

A *simple game* is a tuple $(N, \mathcal{W})$ where N is a finite set of players and $\mathcal{W}$ is a monotonic family of subsets of N formed by the *winning coalitions*, such that if $X \in \mathcal{W}$ and $X \subseteq Z$, then $Z \in \mathcal{W}$. An *influence game* is a simple game defined by a tuple (G, w, f, q) where (G, w, f) is an influence graph and q is a *quota* $0 \leq q \leq |V| + 1$. $X \subseteq V$ is a *winning coalition* iff $|F(X)| \geq q$, otherwise X is a *losing coalition*. Note that every simple game is an influence game [10]. From now on, we assume $N = V$ and $n = |N| = |V|$.

2 Power Indices and New Centrality Measures

A *power index* is a measure of the relevance of the players in a game [1,5]. We consider the two main power indices of a given simple game $(N, \mathcal{W})$. The *Banzhaf index* $\texttt{Bz}(i) = |C_i|/\sum_{i \in N}|C_i|$ and the *Shapley-Shubik index* $\texttt{SS}(i) = (\sum_{S \in C_i}(|S|-1)!\,(n-|S|)!)/n!$, where $C_i = \{S \in \mathcal{W} \mid S \backslash \{i\} \notin \mathcal{W}\}$.

Power indices, in influence games, can be considered centrality measures because an actor is more central in a network while more necessary is for generating of winning coalitions. Moreover, influence games also provide other *new* criteria to determine measures of centrality. Let $f(S) = \sum_{i \in S} f(i)$, for a coalition $S \subseteq N$. For an influence game (G, w, f, q), $\texttt{Effort}(i) = \min\{f(S) \mid |F(S \cup \{i\})| \geq q\}$, the (minimum) effort required by the network to choose a winning coalition that

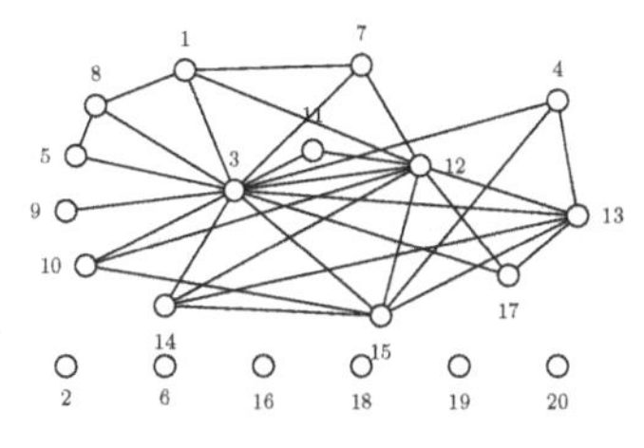

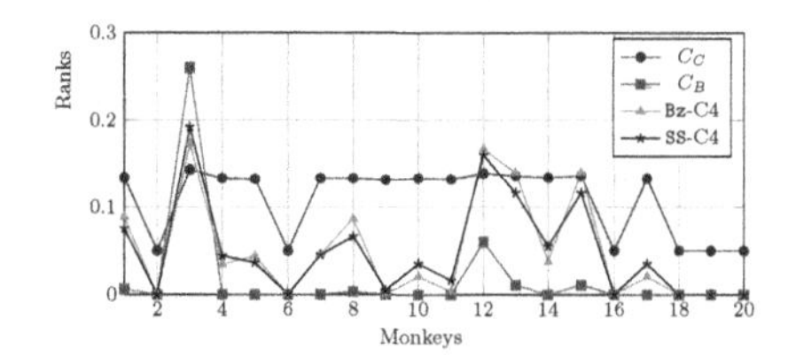

Fig. 1. Monkeys' interaction network and comparisons among Bz-C4, SS-C4, C_C, C_B

contains a required actor. While greater is the required effort for a node, this node should be less central. Therefore, the *effort centrality measure* is the effort required to make the social network follows the opinion of an actor, i.e., $C_E(i) = (f(N) - \texttt{Effort}(i))/f(N)$.

The second new measure is the *satisfaction centrality measure*, based on the *satisfaction score* [14], representing the level of satisfaction of each actor applied to an influence game (G, w, f, q), i.e., $C_S(i) = (|\mathcal{W}_i| + |\mathcal{L}_{-i}|)/2^n$, where $\mathcal{W}_i = \{X \subseteq V(G) \mid i \in X, |F(X)| \geq q\}$ and $\mathcal{L}_{-i} = \{X \subseteq V(G) \mid i \notin X, |F(X)| < q\}$.

3 Cases of Study

We consider three simple real social networks to compare the new centrality measures Bz, SS, C_E and C_S, with some traditional ones, C_D, C_C and C_B. In each comparative table the three more central values will be highlighted in bold. We used enough significant digits to distinguish all the different values.

Monkeys' interaction. This is a network representing the real interactions amongst a group of 20 monkeys observed during three months next to a river provided in [3]. It is represented by an undirected graph with an edge $\{i, j\}$ whenever monkeys i and j were witnessed together in the river. See Figure 1, on the left.

In order to analyze this network $((V, E), w)$ we assume, as usual, that every undirected edge $\{i, j\}$ with $i, j \in V$ represents in fact two arcs (i, j) and (j, i) of E, and the weight function is defined by $w(e) = 1$, for all $e \in E$. In our context, this means that a monkey can influence and be influenced by other monkey if and only if they have interacted. To define an influence game we have to set the quota and define the labeling function. We select $q = 14$, which corresponds to the maximum spread of influence which can be obtained from a monkey. We consider four labeling functions representing different influence requirements. For every node $i \in V$, (C1) minimum, $f(i) = 1$; (C2) average, $f(i) = \lceil deg(i)/2 \rceil$; (C3) majority, $f(i) = \lfloor deg(i)/2 \rfloor + 1$; and (C4) maximum, $f(i) = deg(i)$.

The Bz, SS, C_E and C_S measures have been computed for all these cases (Table 1). Note that only isolated nodes for Bz, SS and C_E, as well as the last column of C_E assume a score exactly equal to zero. For (C1), the new measures

Table 1. Comparison for the Monkeys' interaction network for $q = 14$

				Bz				SS				C_E				C_S			
Node	C_D	C_C	C_B	C1	C2	C3	C4	C1	C2	C3	C4	C1	C2	C3	C4	C1	C2	C3	C4
1	0.21	0.134	0.006	0.07	0.038	0.0708	0.0885	0.07	0.025	0.068	0.075	0.9	0.43	0.14	0	0.501	0.521	0.575	0.598
2	0.00	0.050	0.000	0.00	0.000	0.0000	0.0000	0.00	0.000	0.000	0.000	0.0	0.00	0.00	0	0.500	0.500	0.500	0.500
3	**0.68**	**0.143**	**0.260**	0.07	**0.156**	**0.1214**	**0.1730**	0.07	**0.219**	**0.150**	**0.192**	0.9	0.36	0.07	0	0.501	**0.589**	**0.644**	**0.736**
4	0.16	0.133	0.000	0.07	0.059	0.0673	0.0343	0.07	0.047	0.062	0.044	0.9	**0.50**	0.14	0	0.501	0.537	0.547	0.580
5	0.11	0.132	0.000	0.07	0.019	0.0373	0.0438	0.07	0.013	0.032	0.036	0.9	0.43	0.14	0	0.501	0.510	0.543	0.578
6	0.00	0.050	0.000	0.00	0.000	0.0000	0.0000	0.00	0.000	0.000	0.000	0.0	0.00	0.00	0	0.500	0.500	0.500	0.500
7	0.16	0.133	0.000	0.07	0.049	0.0497	0.0460	0.07	0.032	0.043	0.045	0.9	0.43	0.14	0	0.501	0.528	0.551	0.583
8	0.16	0.133	0.003	0.07	0.048	0.0282	0.0863	0.07	0.040	0.024	0.066	0.9	0.43	0.07	0	0.501	0.527	0.532	**0.601**
9	0.05	0.131	0.000	0.07	0.028	0.0281	0.0003	0.07	0.017	0.022	0.005	0.9	0.43	0.14	0	0.501	0.516	0.531	0.548
10	0.16	0.133	0.000	0.07	0.074	0.0538	0.0205	0.07	0.069	0.050	0.035	0.9	**0.50**	0.14	0	0.501	0.536	0.555	0.582
11	0.11	0.132	0.000	0.07	0.037	0.0470	0.0035	0.07	0.023	0.040	0.016	0.9	**0.50**	0.14	0	0.501	0.520	0.553	0.574
12	**0.47**	**0.139**	**0.060**	0.07	**0.154**	0.1004	**0.1671**	0.07	**0.180**	**0.107**	**0.160**	0.9	0.43	0.14	0	0.501	**0.580**	**0.604**	**0.625**
13	**0.32**	**0.136**	**0.011**	0.07	**0.091**	**0.1197**	**0.1395**	0.07	**0.096**	**0.125**	**0.116**	0.9	0.43	0.07	0	0.501	**0.546**	**0.586**	0.596
14	0.21	0.134	0.000	0.07	0.081	**0.1028**	0.0375	0.07	0.075	0.100	0.055	0.9	**0.50**	0.14	0	0.501	0.541	0.569	0.584
15	**0.32**	**0.136**	**0.011**	0.07	**0.091**	**0.1197**	**0.1395**	0.07	**0.096**	**0.125**	**0.116**	0.9	0.43	0.07	0	0.501	**0.546**	**0.586**	0.596
16	0.00	0.050	0.000	0.00	0.000	0.0000	0.0000	0.00	0.000	0.000	0.000	0.0	0.00	0.00	0	0.500	0.500	0.500	0.500
17	0.16	0.133	0.000	0.07	0.074	0.0538	0.0205	0.07	0.069	0.050	0.035	0.9	**0.50**	0.14	0	0.501	0.536	0.555	0.582
18	0.00	0.050	0.000	0.00	0.000	0.0000	0.0000	0.00	0.000	0.000	0.000	0.0	0.00	0.00	0	0.500	0.500	0.500	0.500
19	0.00	0.050	0.000	0.00	0.000	0.0000	0.0000	0.00	0.000	0.000	0.000	0.0	0.00	0.00	0	0.500	0.500	0.500	0.500
20	0.00	0.050	0.000	0.00	0.000	0.0000	0.0000	0.00	0.000	0.000	0.000	0.0	0.00	0.00	0	0.500	0.500	0.500	0.500

are not good representatives. As the spread of influence is fluid, i.e., actors do not require too many restrictions to form winning coalitions, then all the non-isolated nodes have the same value. However, for the other cases in which differences between influence are relevant, only the pair of monkeys $(10, 17)$ and $(13, 15)$ assume the same value for Bz, SS and C_S, allowing a more relevant classification.

Dining-table partners. This network represents the companion preferences of 26 girls living in one cottage at a New York state training school [11,2]. See Figure 2, on left. Each girl was asked about who prefers as dining-table partner in first and second place. Thus, each girl is represented by a node, and there is a directed edge (i, j) per each girl i preferring girl j as dining-table partner. Every node has an outdegree equal to 2: edges with weight 1 denote the first option of the girl, and edges with weight 2 denote her second option.

We could assume that a girl has some ability to influence another one which has chosen her as a partner. Figure 2 (on right) shows the corresponding network of this influence game, reversing each arc (i, j) by (j, i), so that a node points to another when the first one has some influence over the second one. Further, the weights of the edges must be exchanged, so that an original edge (i, j) with weight 1 now becomes in an edge (j, i) with weight 2, and viceversa. Because a girl has more influence over another one if that other has chosen her in the first place rather than in the second place. Of course, now every node has an indegree equal to 2: one edge with weight 1 and the other with weight 2. We consider a quota $q = 14$, so that a coalition is winning if and only if it achieves to convince (through its spread of influence) most of the girls *absolute majority.* For every node $i \in V$, we consider the following labeling functions:

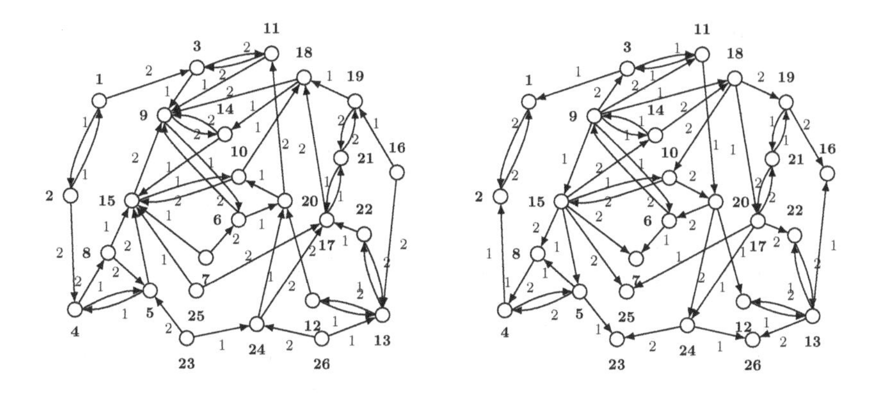

Fig. 2. Network for dining-table partners and the associated influence graph

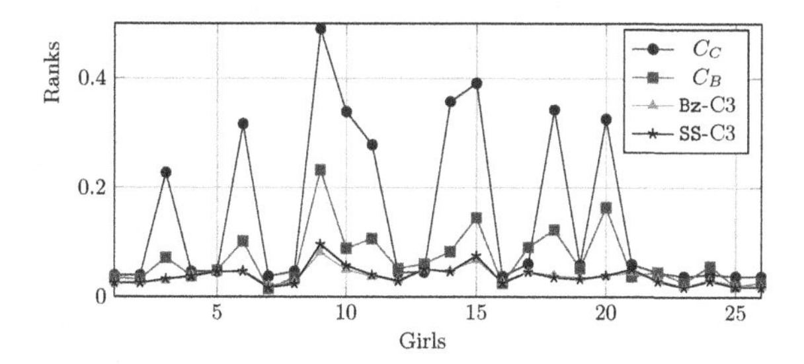

Fig. 3. Comparative between Bz-C3, SS-C3, C_C and C_B for Dining-table partners

(C1) minimum, $f(i) = 1$; (C2) average, $f(i) = 2$; and (C3) maximum, $f(i) = 3$. Unlike in the previous network, here there are no isolated nodes, but we can still obtaining scores for Bz and SS equal to zero. See the columns of Bz-C1 and SS-C1 on Table 2.

Indegree centrality C_D^- does not provide any relevant information, because the indegree for each node is always 2 (see Table 2 and Figure 3). Similarly as it succeded in the previous network, Bz-C1, SS-C1, C_E-C1 and C_S-C1 have several nodes with the same rank, but while the required influence to convincement increases, the values of the measures are more diverse for the power indices and satisfaction centrality. Measures Bz-C2, SS-C2 and C_S-C2, as well as C_D^+ and C_C, have only some values that are repeated, but measures Bz-C3, SS-C3 and C_S-C3 have the same values only for girls 1 and 2. These girls are equivalent in this sense for all the other measures except by C_B, in which, however, together with C_E, girls 23 and 26 have the same centrality.

Girl 15 has a high centrality in all measures, as well as girl 9, except in C_E-C2, as well as in Bz-C2 and SS-C2, where is far less central. Girl 13 is fairly central exclusively in C_D^+, because despite of its high outdegree, only exist paths from this node to another four, which is a severe restriction for all other measures.

Table 2. Comparison values for the Dining-table partners network for $q = 14$

					Bz			SS			C_E			C_S		
Node	C_D^-	C_D^+	C_C	C_B	C1	C2	C3	C1	C2	C3	C1	C2	C3	C1	C2	C3
1	0.08	0.04	0.0400	0.035	0.00	0.028	0.0274	0.0000	0.0103	0.0259	0.92	0.85	0.42	0.500000	0.5024	0.5300
2	0.08	0.04	0.0400	0.033	0.00	0.028	0.0274	0.0000	0.0103	0.0259	0.92	0.85	0.42	0.500000	0.5024	0.5300
3	0.08	0.08	0.2273	0.072	**0.08**	0.008	0.0302	**0.0832**	0.0014	0.0331	**0.96**	0.85	**0.54**	**0.500217**	0.5006	0.5329
4	0.08	0.08	0.0473	0.039	0.00	0.028	0.0413	0.0000	0.0103	0.0383	0.92	0.85	0.42	0.500000	0.5024	0.5451
5	0.08	0.12	0.0473	0.049	0.00	0.028	0.0452	0.0000	0.0103	0.0463	0.92	0.85	0.42	0.500000	0.5024	0.5494
6	0.08	0.08	0.3165	0.102	**0.08**	0.043	0.0481	**0.0832**	0.0142	0.0473	**0.96**	0.85	0.42	**0.500217**	0.5036	0.5526
7	0.08	0.00	0.0385	0.015	0.01	0.024	0.0216	0.0003	0.0075	0.0176	0.92	0.85	0.42	0.500027	0.5020	0.5236
8	0.08	0.04	0.0471	0.036	0.00	0.024	0.0278	0.0000	0.0075	0.0239	0.92	0.85	0.42	0.500000	0.5020	0.5303
9	0.08	**0.24**	**0.4902**	**0.232**	**0.08**	0.027	**0.0820**	**0.0832**	0.0072	**0.0965**	**0.96**	0.85	**0.54**	**0.500217**	0.5023	**0.5896**
10	0.08	0.08	0.3378	0.089	**0.08**	**0.104**	**0.0506**	**0.0832**	**0.1953**	**0.0578**	**0.96**	**0.92**	**0.54**	**0.500217**	**0.5089**	**0.5552**
11	0.08	0.08	0.2778	0.107	**0.08**	0.008	0.0383	**0.0832**	0.0014	0.0410	**0.96**	0.85	**0.54**	**0.500217**	0.5006	0.5418
12	0.08	0.04	0.0452	0.052	0.00	0.004	0.0321	0.0001	0.0007	0.0292	0.92	0.85	0.42	0.500004	0.5003	0.5350
13	0.08	**0.16**	0.0454	0.061	0.00	0.014	0.0500	0.0001	0.0041	0.0511	0.92	0.85	0.42	0.500004	0.5012	0.5546
14	0.08	0.08	**0.3571**	0.083	**0.08**	**0.104**	0.0486	**0.0832**	**0.1953**	0.0465	**0.96**	**0.92**	0.42	**0.500217**	**0.5089**	0.5531
15	0.08	**0.24**	**0.3906**	**0.145**	**0.08**	**0.104**	**0.0683**	**0.0832**	**0.1953**	**0.0755**	**0.96**	**0.92**	**0.54**	**0.500217**	**0.5089**	**0.5746**
16	0.08	0.00	0.0385	0.025	0.00	0.015	0.0279	0.0000	0.0055	0.0256	0.92	0.85	0.42	0.500000	0.5013	0.5305
17	0.08	**0.33**	0.0614	0.091	**0.08**	**0.051**	0.0469	**0.0832**	**0.0232**	0.0459	**0.96**	0.85	0.42	**0.500217**	**0.5043**	0.5512
18	0.08	0.12	0.3425	0.123	**0.08**	**0.104**	0.0404	**0.0832**	**0.1953**	0.0361	**0.96**	**0.92**	0.42	**0.500217**	**0.5089**	0.5442
19	0.08	0.08	0.0595	0.053	**0.08**	0.036	0.0356	**0.0832**	0.0126	0.0327	**0.96**	0.85	0.42	**0.500217**	0.5031	0.5389
20	0.08	0.12	0.3247	**0.164**	**0.08**	**0.075**	0.0394	**0.0832**	**0.0413**	0.0395	**0.96**	0.85	0.42	**0.500217**	**0.5064**	0.5430
21	0.08	0.08	0.0605	0.038	**0.08**	**0.051**	0.0457	**0.0832**	**0.0232**	0.0512	**0.96**	0.85	**0.54**	**0.500217**	**0.5043**	0.5499
22	0.08	0.04	0.0452	0.046	0.00	0.025	0.0325	0.0001	0.0082	0.0293	0.92	0.85	0.42	0.500004	0.5021	0.5355
23	0.08	0.00	0.0385	0.027	0.00	0.011	0.0191	0.0000	0.0029	0.0173	0.92	0.85	0.42	0.500000	0.5010	0.5208
24	0.08	0.08	0.0417	0.057	0.01	0.029	0.0324	0.0003	0.0083	0.0301	0.92	0.85	0.42	0.500027	0.5025	0.5354
25	0.08	0.00	0.0385	0.020	0.01	0.024	0.0218	0.0003	0.0075	0.0187	0.92	0.85	0.42	0.500027	0.5020	0.5239
26	0.08	0.00	0.0385	0.027	0.00	0.004	0.0197	0.0000	0.0007	0.0177	0.92	0.85	0.42	0.500000	0.5003	0.5215

Student Government discussion. The last case of study starts with the social network illustrated in Figure 4. This network represents the communication interactions among different members of the Student Government at the University of Ljubljana in Slovenia. Data were collected through personal interviews in 1992 and published by [6], being used later by [2].

Every directed edge is a communication interaction and all of them have the same weight equal to 1. Each node is a member of the Student Government, and unlike the previous cases, here nodes are labeled beforehand: There are three *advisors* labeled 1, seven *ministers* labeled 2, and one *prime minister* labeled 3.

We modified slightly this network to obtain the influence graph of Figure 4. We assume that every communication interaction is an attempt to influence another student. Thus, the capacity to influence depends on the student's position. For instance, the advise of a prime minister does not have the same effectiveness —marked with weight 3— than the advise of an advisor —marked with weight 1. Furthermore, as the labels of the nodes should represent the difficulty of each student $i \in N$ to be influenced, according to their position in the Student Government, then they have been changed by the following values: $f(i) = 1$, if i is an advisor; $f(i) = \lceil deg^-(i)/2 \rceil$, if i is a minister; and $f(i) = deg^-(i)$, if i is the primer minister. We consider a majority influence required to win, setting $q = 6$ (see Table 3 and Figure 4).

Traditional measures provide different rankings. In fact, none of the most central nodes measured with C_C and C_B coincide, and while the most central node for C_C is the advisor 10, this is the less central according to C_B. Moreover, the ministers 3 and 1 are very central for C_C but with C_B are at the bottom of the ranking. This is because nodes 1, 3 and 10 have a high accessibility to all other nodes, but however, they are not good intermediaries for connecting

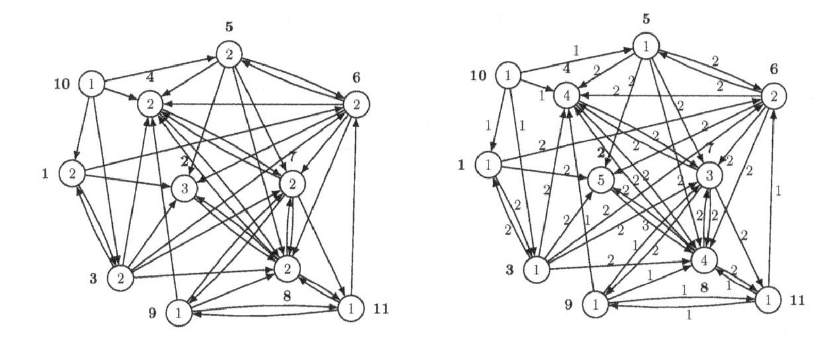

Fig. 4. Student Government discussion network and the asssociated influence graph

Table 3. Comparison for the Student Government discussion networ and $q = 6$

Node	C_D^-	C_D^+	C_C	C_B	Bz	SS	C_E	C_S
1	0.2	0.3	**0.357**	0.130	**0.164**	**0.176**	**0.91**	**0.516**
2	0.5	0.1	0.200	0.195	0.154	0.076	0.45	0.515
3	0.2	**0.6**	**0.435**	0.169	**0.164**	**0.176**	**0.91**	**0.516**
4	**0.7**	0.2	0.208	0.204	0.005	0.009	0.55	0.500
5	0.2	**0.5**	0.238	0.211	**0.164**	**0.176**	**0.91**	**0.516**
6	0.4	**0.5**	0.238	**0.304**	**0.164**	**0.176**	0.82	**0.516**
7	**0.6**	**0.4**	0.227	**0.316**	0.005	0.009	0.64	0.500
8	**0.8**	**0.4**	0.227	0.262	0.005	0.009	0.55	0.500
9	0.2	**0.4**	0.227	0.193	0.005	0.009	0.82	0.500
10	0.0	**0.4**	**0.556**	0.111	**0.164**	**0.176**	**0.91**	**0.516**
11	0.3	0.3	0.227	**0.306**	0.005	0.009	0.82	0.500

distant nodes through paths. Nevertheless, nodes 1, 3 and 10, as well as ministers 5 and 6, have a high score for measures Bz, SS and C_S. Thi is so since the spread of the influence over the other students, starting from the coalitions where they participate, is often necessary to overcome the required quota q. The same occurs for C_E, except for the minister 6, which is a bit less central.

4 Conclusions and Future Work

Our main motivation in this work was to use influence games as a way to propose additional centrality measures coming from the field of cooperative game theory. The framework of influence games derives a connection between social network analysis and spread of influence in decision processes. We exploit this link with simple game theory to propose new centrality measures: Banzhaf, Shapley-Shubik, Effort and Satisfaction. This is the first approach to apply power indices as centrality measures for social networks (for *specific* game-theoretic networks [9] has been used the Shapley-Shubik index as centrality measure). Our results do not contradict the relevance criteria provided by traditional centrality measures like degree centrality, closeness or betweenness. In some cases such measurements are similar to our measurements, but there are also cases where the results have been quite different. Indicating that an additional study on more realistic social networks is of interest.

Our proposal can be extended to other power indices [5] and measures, it will be of interest to determine which of them provide relevant rankings for social network analysis. Finally, we want to mention that there are other well known concepts related with players in simple games, such as *dummy, vetoer* or *dictators* [13], that could provide interesting properties of actors in a social network.

References

1. Aziz, H.: Algorithmic and complexity aspects of simple coalitional games. PhD Thesis, Department of Computer Science, University of Warwick (2009)
2. de Nooy, W., Mrvar, A., Batagelj, V.: Exploratory social network analysis with Pajek. Structural analysis in the social sciences, vol. 27. Cambridge Univ. Press (2005)
3. Everett, M.G., Borgatti, S.P.: The centrality of groups and classes. Journal of Mathematical Sociology 23(3), 181–201 (1999)
4. Freeman, L.C.: Centrality in social networks: Conceptual clarification. Social Networks 1(3), 215–239 (1979)
5. Freixas, J.: Power indices. In: Cochran, J.J., Cox, L.A., Keskinocak, P., Kharoufeh, J.P., Smith, J.C. (eds.) Wiley Encyclopedia of Operations Research and Management Science, vol. 8, John Wiley & Sons (2011)
6. Hlebec, V.: Recall versus recognition: Comparison of two alternative procedures for collecting social network data. In: Ferligoj, A., Kramberger, T. (eds.) Proceedings of the International Conference on Methodology and Statistics, pp. 121–128 (1992, 1993)
7. Kempe, D., Kleinberg, J., Tardos, É.: Maximizing the spread of influence through a social network. In: Getoor, L., Senator, T.E., Domingos, P., Faloutsos, C. (eds.) Proceedings of the Ninth ACM SIGKDD International Conference on Knowledge Discovery and Data Mining, pp. 137–146 (2003)
8. Latora, V., Marchiori, M.: A measure of centrality based on network efficiency. New Journal of Physics 9(6) (2007)
9. Michalak, T.P., Aadithya, K.V., Szczepański, P.L., Ravindran, B., Jennings, N.R.: Efficient computation of the Shapley value for game-theoretic network centrality. Journal of Artificial Intelligence Research 46, 607–650 (2013)
10. Molinero, X., Riquelme, F., Serna, M.J.: Social influence as a voting system: A complexity analysis of parameters and properties. CoRR, abs/1208.3751v3 (2014)
11. Moreno, J.L.: The sociometry reader. Free Press (1960)
12. Sun, J., Tang, J.: A survey of models and algorithms for social influence analysis. In: Aggarwal (ed.) Social Network Data Analytics, pp. 177–214 (2011)
13. Taylor, A., Zwicker, W.: Simple games: Desirability relations, trading, pseudoweightings. Princeton University Press (1999)
14. van den Brink, R., Rusinowska, A., Steffen, F.: Measuring power and satisfaction in societies with opinion leaders: dictator and opinion leader properties. Homo Oeconomicus 28(1-2), 161–185 (2011)
15. Wasserman, S., Faust, K.: Social network analysis: Methods and applications. Structural analysis in the social sciences, vol. 8. Cambridge Univ. Press (1994)

Self-Healing Multi Agent Prototyping System for Crop Production

Haeng-Kon Kim[1] and Hyun Yeo[2]

[1] School of Information Technology, Catholic University of Daegu, Kyungbuk, 712-702, Korea
hangkon@cu.c.kr

[2] Dept. of Information and Communication Engineering, Sunchon National University, Suncheon, Jeollanam-do, Republic of Korea
yhyun@sunchon.ac.kr

Abstract. An agent is a computer system capable of flexible and autonomous action in dynamic, unpredictable and typically multi-agent domains. Most distributed computing environments today are extremely complex and time-consuming for human administrators to manage. Thus, there is increasing demand for the self-healing and self-diagnosing of problems or errors arising in systems operating within today's ubiquitous computing environment. This paper proposes a proactive self-healing system that monitors, diagnoses and heals its own internal problems using self-awareness as contextual information for crop production monitoring system in the future. The proposed system consists of Multi-Agents that analyze the log context, error events and resource status in order to perform self-healing and self-diagnosis. To minimize the resources used by the Adapters which monitor the logs in an existing system, we place a single process in memory. By this, we mean a single Monitoring Agent monitors the context of the logs generated by the different system components. For rapid and efficient self-healing, we use a 6-step process. The effectiveness of the proposed system is confirmed through practical experiments conducted with a prototype system.

Keywords: Self-healing, Self-diagnosing, Agent, Ubiquitous computing, CBE(Common Base Event), Crop Production Agent Systems.

1 Introduction

Applications where agent technologies will play a crucial role include: Ambient Intelligence, Grid Computing, e-Business, the Semantic Web, and Bio-informatics. Agent systems are one of the most vibrant and important areas of research and development to have emerged in information technology. Most distributed computing environments today are very complex and time consuming for a human administrator to manage. Thus, there is a growing need for experts who can assure the efficient management of various computer systems. However, management operations involving human intervention have clear limits in terms of cost effectiveness and the availability

C. Ramos et al. (eds.), *Ambient Intelligence - Software and Applications*,
Advances in Intelligent Systems and Computing 291,
DOI: 10.1007/978-3-319-07596-9_4, © Springer International Publishing Switzerland 2014

of human resources [1]. With regard to all computer problems, about 40% are attributable to errors made by the system administrators [2]. Thus, the current system management method, which depends mainly on professional managers, must be improved.

In ubiquitous environments, which involve an even greater number of computing devices, with more informal modes of operation, this type of problem will have rather serious consequences. In order to solve these problems when they arise, effective self-healing systems are required. A self-healing system allows the system or computing device itself to recognize, identify and heal problems arising, without depending on administrators [3].

The existing self-healing systems consist of a 5-step process, including Monitoring, Translation, Analysis, Diagnosis and Feedback. This architecture has various drawbacks. These drawbacks are presented as the existing system is a log-based system, if an error or problem arising in a component does not generate a log event, it can't heal the problem or error. It also increased log file sizes and frequency and the wastage of resources (such as 'RAM', 'CPU', etc). It is a lot of dependency on the administrator and vendor.

In this paper, we propose an SHMAP (Self-Healing Multi Agent Prototyping proactive self-healing system which incorporates several functions designed to resolve the problems mentioned above, namely **(1)** the minimization of the resources required through the use of a single process (Monitoring Agent), **(2)** the use of a Meta Policy which offers different healing strategies according to the components situation, viz. 'Emergency', 'Alert', 'Error' and 'Warn'**.** **(3)** for the sake of rapid and efficient self-healing, we use a 6-step process. The proposed system is designed and implemented in the form of a prototype, in order to prove its effectiveness through experimentation.

2 Related Works

Self-adaptive software has the capability to modify its own actions in response to changes in areas such as system faults, and resource variability. These self-adaptive behaviors are the essence of the self-healing system or, in other words, the self-healing system must contain self-adaptive behaviors [11]. Oreizy et. al. [4] proposed the following processes for self-adaptive software: Monitoring the system, Planning changes, Deploying the change descriptions and Enacting these changes [5][6][7][8].

2.1 Adaptive Service Framework (ASF)

The Adaptive Service Framework (ASF) [10] proposed by IBM and CISCO is applied in the form of self-adaptive behaviors. The functions of the ASF are the *Adapters* [9] monitor the logs from the various components and the *Adapter* translates the logs generated by the component into the CBE (Common Based Event) format.

The *Autonomic Manager* [9] analyzes the CBE log. This step identifies the relationship between the components through their dependency. The *Autonomic Manager* [9] finds the appropriate healing method by means of the *Symptom Rule* [9] and *Policy Engine* [9] then applies the healing method to the applicable component. The feedback from the *Resource Manager* [9] enables the system to heal itself.

In the event that the component has a critical problem or one which cannot be solved easily, the Autonomic Manager sends a *Call Home Format* message to the Support Service Provider (SSP) / Vendor, requesting them to find a solution.

2.2 Context Awareness for Self-adaptive Software

Adaptive software evaluates its own action. If it does not perform an intended goal, it changes its own action in order to achieve more efficient performance [14]. It must consider a realistic environment for system operation, and possess appropriate architecture operating in the external environment. Through the architecture, the self-adaptive software is able to change its own behavior to achieve an intended goal. These behavior changes achieve results depending on the context in the environment. It is difficult to adapt in changing external environments using the contextual information, because situation information is widely variable and representation is not accurately defined [15]. To develop Adaptive software, the software needs to involve context-awareness for external environments. The environment during operating software is called 'context'.

However, it has progressed slowly in the study of context self-awareness. Thus, this paper proposes a 'proactive self-healing system' that knows the situation in system and infers when system failure occurs.

3 Design of SHMAP

3.1 Basic Designs Concepts

The functions of the proposed SHMAP system are as follows:

(1) the Monitoring Agent, which runs as a single process, provides real time monitoring of error events, in order to overcome the problem associated with the number of *Adaptors* [10], which result in memory wastage. (2) if problems or errors arising in systems result in an emergency situation, the Component Agent can take proactive and immediate action. (3) the System Agent using threshold values of the system resources recognizes system situations and chooses an applicable policy on the Meta level. (4) before translating the original log into the log of the CBE type, a filtering process is performed. In effect, the Component Agent reduces the size of the log file. (5) in the event that there is no applicable method of healing, the Searching Agent searches the web server of the vendor in order to interact with the administrator.

3.2 System Architecture

Fig. 1 shows the structure of the proposed system, which is composed of a Monitoring Agent, Component Agent, System Agent, Diagnosis Agent, Decision Agent and Searching Agent. The Proposed system consists of 6 consecutive processes, viz. Monitoring, Filtering, Translation, Analysis, Diagnosis and Decision and Feedback

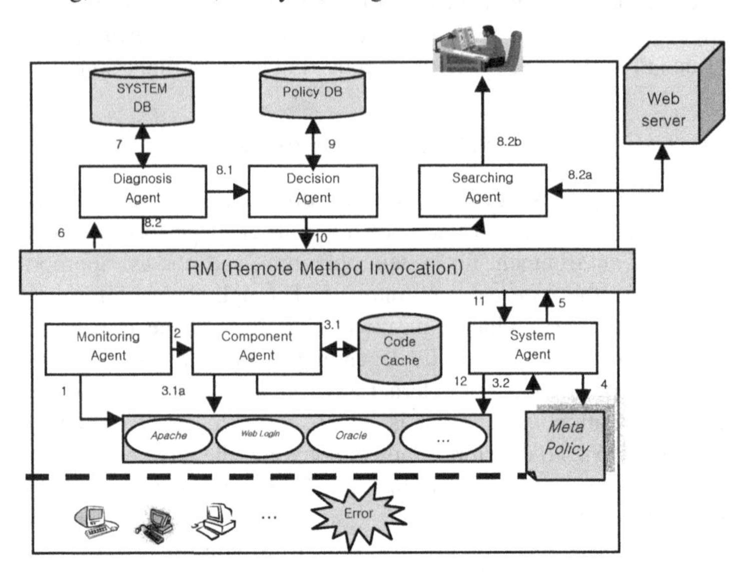

Fig. 1. Architecture of SHMAP Systems

3.2.1 Monitoring Agent (*Monitoring Phase*)

As shown in Fig.2, the functions of the Monitoring Agent are as follows:

It monitors resource (such as Ram, CPU, etc) status and the size of the log file generated by the components. To deal with errors or problems arising in the component that do not generate log events, it monitors error events arising in the operating system, in order to detect problems or errors concerning these components.

Through resource status, log files and error events, if the Monitoring Agent detects suspicious events of the components, it executes the Component Agent.

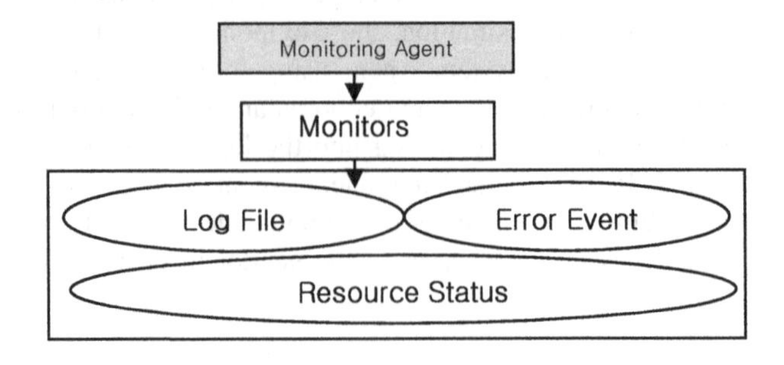

Fig. 2. Monitoring Agent's behavior

3.2.2 Component Agent (Filtering and Translation Phase)

The functions of the Component Agent are as follows; It gathers the information provided by the Monitoring Agent. It filters error context from normal log context File (as shown in Fig 3). As can be seen, the error context is filtered by means of designated keywords, such as "*not*", "*error*", "*reject*", "*notify*", etc. It also translates the filtered error context into the CBE format. Using the syslog service existing in the operating system, as shown in Table1, it classifies the Error Event, and sets up the priorities. It takes immediate action corresponding to Emergency (Priority '1') situation, through the code cache

```
bufferedReader br = ReadeLineFromFile(logfile);
while((Line = br.readLine()|=null)(
      if((Line.indexOf("not") > 0)) {
                Line = Line + "ERROR"+"₩n";
      }
      else if((Line.indexOf("error") > 0)) {
                Line = Line + "ERROR"+"₩n";
      }
      else if((Line.indexOf("reject") > 0)) {
                Line = Line + "Error"+"₩n";
      }
                ...
      else
                Line = Line + "NORMAL"+"₩n";
      log = log + Line;
}
WriteLineToFile(log.parsedfile);
...
```

Fig. 3. The filtering source code

Table 1. Classification of the Error Event

Error Level	Priority
Emergency	1
Alert	2
Error	3
Warn	4

The log contexts generated by the component as the result of observing an error is referred to as an error report. The Component Agent can take proactive and immediate action. As shown in Fig.4, the Component Agent generates the Error Report and the CBE. The Error Report is an administrator document, and the CBE is a document for the system.

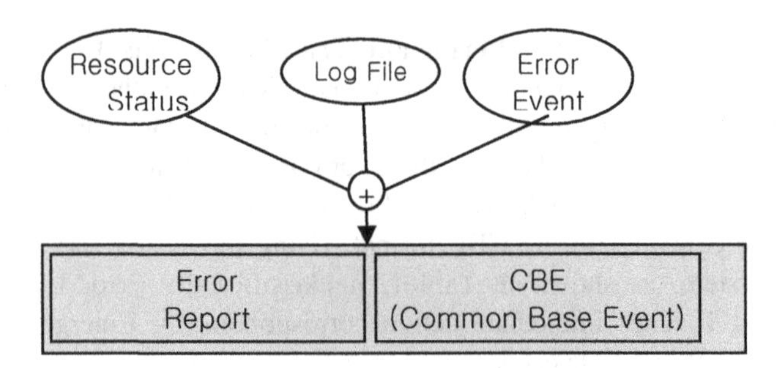

Fig. 4. Component Agent's behavior

3.2.3 System Agent (*Execution Phase*)

The System Agent consists of the CBE Log Receiver, Resource Collector, Adaptation Module and Executor, as shown in Fig 5. When the System Agent receives the CBE log from the CBE Log Receiver, the Resource Collector gathers the CPU information, Memory information, process information and Job Schedule information, in order to deliver it to the Adaptation Module.

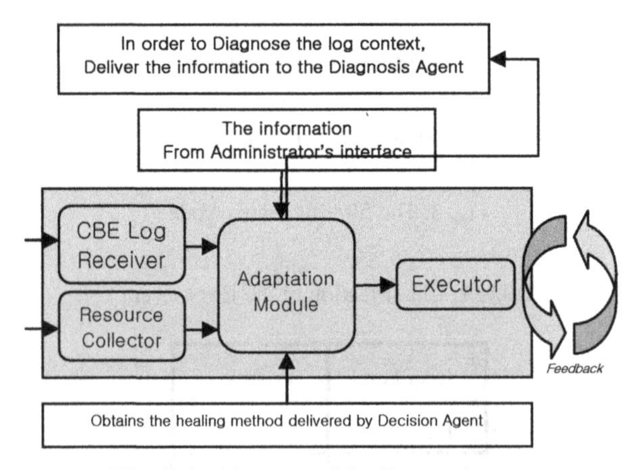

Fig. 5. Architecture of the System Agent

The Adaptation Module is in possession of the threshold values pertaining to the gathered resource information. According to the threshold value, a suitable policy is implemented. The Executor then executes the best healing method. In this step, we distinguish the dependency of the components. The collected information and CBE logs are sent to the Diagnosis Agent. The System Agent applies the best healing method in accordance with the threshold value.

3.2.4 Diagnosis Agent (Analysis and Diagnosis Phase)

The first major attribute of a self-healing system is self-diagnosing [9]. The Diagnosis Agent analyzes the CBE log, resource information (received from the System Agent) and the dependency of the components, and then diagnoses the current problem (through the Symptom DB). It provides technology to automatically diagnose problems from observed symptoms. The results of the diagnosis can be used to trigger automated reactions.

The following is the algorithm performed by the Diagnosis Agent.
Step1 Extract the CBE log from an encapsulated message
Step2 Read the correlated component in the CBE log
Step3 Search the correlated CBE logs during the assigned time
Step4 Cluster the correlated logs
Step5 Diagnose a set of the logs using IF~THEN rules
Step6 Send a result to the Decision Agent with additional Information

3.2.5 Decision Agent (Decision and Feedback Phase)

Through the information delivered by the Diagnosis Agent, The Decision Agent determines the appropriate healing method with the help of the *Policy DB*. It also receives feedback information from the System Agent in order to apply the more efficient healing method.

The Information received from the diagnosis Agent is used to determine the healing method. The Decision Agent determines the solutions that can be classified into root healing, temporary healing, first temporary healing and second root healing.

Temporary healing is a way of resolving a problem temporarily, such as disconnecting a network connection, assigning temporary memory. The root healing is the fundamental solutions on the diagnosed result, including re-setting, restarting, and rebooting. The Decision Agent stores the methods in the DB as below Fig. 6, and decides how to select the appropriate healing method. However, in the event that the desired conditions do not exist, using the Decision Tree, the Decision Agent determines the most appropriate healing method and sends out the code to heal the system.

Table 2. Decision Table

PROBLEM	TEMPRARY SOULTION	ROOT SOLUTION	CURRENT JOB	FUTURE JOB	AVAILABLE MEMORY	DECISION	FEEDBACK
CONNECT	CHANGE CONFIGURATION	REINSTALL	NULL	NULL	60%	T	POSITIVE
AVAILABLE	NULL	RESTART	NULL	NULL	80%	R	POSITIVE
OVERFLOW	ALLOCATE MEMORY	REBOOT	BACKUP	NULL	30%	TR	POSITIVE

The Table 2 is the table to determine the optimal resolution method by analyzing given attributes. Looking at the DECISION column, when placed under the current diverse context, it helps to determine R(Root Solution), T(Temporary Solution), or TR(first Temporary Solution, second Root Solution). The FEEDBACK Column is showing feedbacks that were executed by the System Agent to heal the system.

The Decision Agent compares the fields with the information received by the System Agent, these fields are CURRENT JOB, FUTURE JOB and AVAILABLE

MEMORY. If the value of the FEEDBACK Column is POSITIVE, the appropriate method is determined.

However, if there are no matched fields, the Decision Tree using the ID3 algorithm generates the generalized classification rules on the given pattern, and can be used in classifying patterns not already in possession.

If the final decision is made through this algorithm, the Decision Agent sends the scripts and codes for healing to the System Agent. At this time, the Decision Agent delivers the appropriate method and several alternative methods assigned with a priority. And then, this learns using reinforcement learning

3.2.6 Searching Agent

The Searching Agent is used to search the vendor's website for the knowledge required to solve the problem, This Agent uses search engines such as *Google* to search for the address of the web site at which the patch specified in the CBE Log can be obtained. It sends the resulting search information to the administrator. The Code Cache is used to provide healing code to solve the error of the component arising in *emergency situations*. If the system may not perform self-healing processes, the administrator interface helps the Administrator heal his system through scripting. It provides a browsing interface to see the information sent by the Searching Agent. The administrator can perform patching or any other action such as scripting or reconfiguring specific files through the interface.

3.3 Meta Policy for Self-Healing

In this section, we describe the *Meta Policy* that reconfigures the component of the proposed system. Using the *Meta Policy*, agents can select the appropriate adaptation policy. Fig. 6 shows that a suitable strategy is selected via a Meta policy, and then the architecture of the proposed system is reconfigured by this strategy.

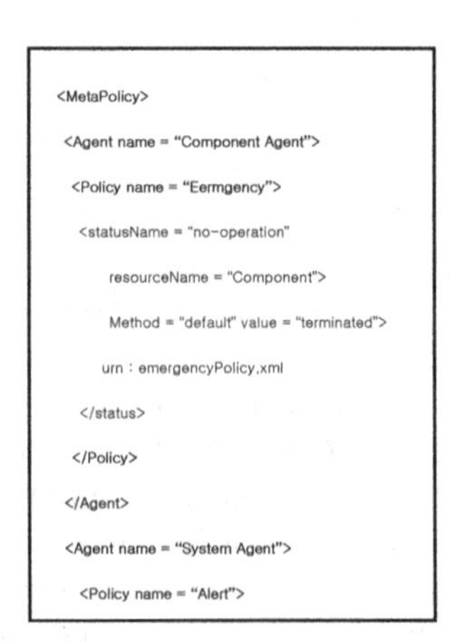

Fig. 6. The Meta Policy of the Component Agent and System Agent

The Meta policy document identifies the different polices that can be applied, namely the "*Emergency*", "*Alert*", "*Error*" and "*Warn*" policies. Theses policies each have a status, linked to their respective policy, and then the architecture of the proposed system is reconfigured by this strategy.

We can understand their behavior from the above document. The Component Agent looks up the current status in the MetaPolicy.xml file. If the agent classifies the current situation as an emergency based on the information contained in the MetaPolicy.xml file, it acts to heal the component which is controlled 표 this agent. The operation that it performs is described in the emergencyPolicy.xml file, in the form of Component Agent behavior.

The System Agent uses the Meta policy to identify the current situation of the system. For example, when the number of processes in memory is 70, the system agent interprets the current situation as an emergency. When the agents classify the status as an emergency, it applies the policy corresponding to the link in the Meta policy document: "urn:<finename.xml>".

4 Implementation and Evaluation of SHMAP

We employed JAVA SDK1.4, and used Oracle9i as the DBMS. Also we used JADE1.3 for the Agent development. The sample log used for the self-healing process was a log generated with *APACHE*. We implemented *the Agents* of the proposed system (in the form of a JADE Agent Platform [13]). As shown in Fig 9 and 10 [13], each of the agents is registered with each of the containers, and the ACL (Agent Communication Language) is used to communicate among the agents. We performed the simulation using six agents.

The Fig.7 shows that the result of the Monitoring Agent. The Fig.8 and Fig.9 illustrate data to be translated to the filtered log and the CBE log, acting as the common log format.

```
The size of FILE 3507bytes
The command that can execute the agent:
Java jade.Boot –host pjm –container
                    sender:SHS_ComponentAgent.java
The location of log:
D:₩₩RELATED_WORK₩₩AUTONOMIC₩₩PROJECT₩₩
Self-healing-
```

Fig. 7. The result of the Monitoring

```
date : [Fri Mar 17 11:32:42 2005]
log_level : [Alert]
client_ip : [client 203.252.53.142]
log_description :
httpd: Could not determine the server's fully
qualified domain name
```

Fig. 8. The filtered log('Alert Situation')

```
<affectedComponentID componentAddressType="HostAddressType">
    <componentData InstanceID="A001" application="Apache" executionEnvironment="RedHatLinux">
    </componentData>
    <componentType name="web_Application_Server">
    </componentType>
    <componentAddress HostAddressType="pjn">
    </componentAddress>
</affectedComponentId>

<reporterComponentId>
    <componentData InstanceID="A001" application="Apache" executionEnvironment="RedHatLinux">
    </componentData>
    <componentType name="web_Application_Server">
    </componentType>
    <componentAddress HostAddressType="pjn">
    </componentAddress>
</reporterComponentId>
```


Fig. 9. CBE (Common Base Event) log

The proposed system was evaluated and compared qualitatively and quantitatively in terms of the Log Monitoring Module, the Filtering & Translating Efficiency, and the healing Time.

(a) Log Monitoring Test. In the existing system, if the number of components is Ω, the system has to have Ω processes to monitor the log. In the proposed system, however, only one process is needed to monitor the log, as shown in Fig. 14. In this figure, the proposed system demonstrates its ability to stay at a certain level of memory usage, even when the number of components in increased.

(b) Filtering & Translation Efficiency Test. In the proposed system, the Component Agent searches for a designated keyword (such as "not", "reject", "fail", "error", etc.) in the log generated by the components. By using this approach, we were able to increase the efficiency of the system, in terms of the size of the log and the number of logs. We analyzed up to 500 logs, filtered out those logs not requiring any action to be taken, and evaluated the number and size of the logs in the case of both the existing and proposed systems. As a result of the filtering process, only about 20% of the logs were required for the healing process, as shown in Fig. 10. Therefore, the proposed system reduces the number and size of the logs, which require conversion to the CBE format.

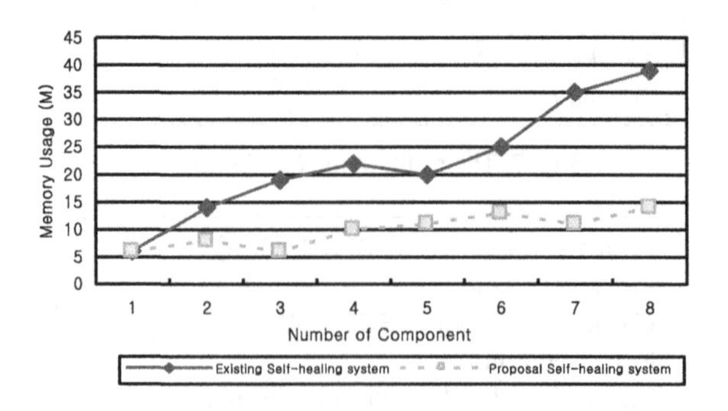

Fig. 10. Memory Usage

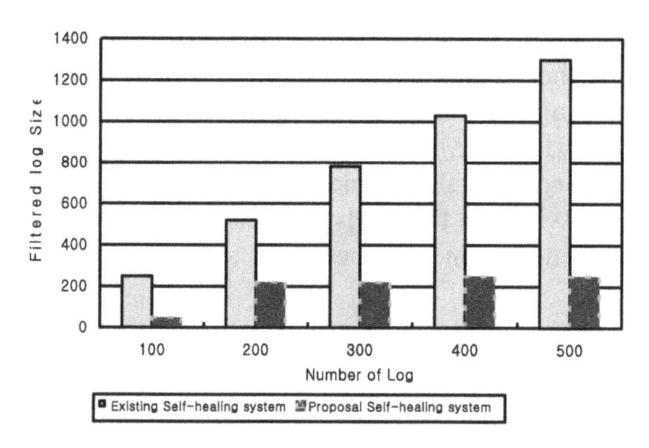

Fig. 11. Comparison of size and number of logs

(c) Average Healing Time Measurement. We measured the Average Healing Time arising in the existing self-healing system and the proposed self-healing system.

We classified the type of error, and measured the average healing time of the classified errors. As shown in Fig. 12, we verified that the proposed systm's healing time is faster than the existing system's healing time and rapidly responded to problems arising in the urgent situation. In the event that the error component does not generate a log, we couldn't measure the healing time arising in the existing self-healing system because the existing system was a log-based healing system.

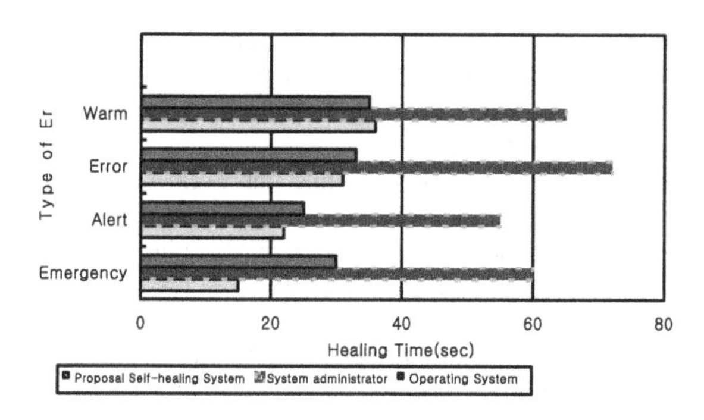

Fig. 12. Comparison of the Healing Time

5 Conclusion

In this paper, we proposed a Multi-Agent based self-healing system, with a pupose of enabling a computer system to observe, diagnose and heal errors or problems in ubiquitous environments. The advantages of this system are as follows. First, when prompt responses are required, the system can make an immediate decision and respond right away. Second, the Monitoring Agent monitors error event and resource status the generation of the log on the fly, thus improving the memory usage. Third, before converting the log into the CBE (Common Base Event) format, filtering is performed to minimize the memory and disk space used in the log conversion. Fourth, it provides a faster healing time. Fifth, using the Meta Policy, the appropriate adaptation policy is selected. However, further study is required to develop an algorithm able to make accurate and general inferences when prompt based responses are employed in the proposed system in this study. It is also necessary to examine the problems associated with the dependency on the vendor and manager.

Acknowledgment. "This research was supported by the MSIP(Ministry of Science, ICT and Future Planning), Korea, under the CITRC(Convergence Information Technology Research Center) support program (NIPA-2013-H0401-13-2008) supervised by the NIPA(National IT Industry Promotion Agency)".

"This research was also supported by the International Research & Development Program of the National Research Foundation of Korea (NRF) funded by the Ministry of Science, ICT & Future Planning(Grant number: K 2012057499)".

References

1. Madarasz, L., Timko, M., Racek, M.: Enterprise Modeling and its Applications in Company Management Systems. In: 5th International Symposium of Hungarian Researchers on Computational Intelligence, Budapest, November 11-12 (2004)
2. Paolucci, M., Sacile, R.: Agent-Based Manufacturing and Control Systems, New Agile Manufacturing Solutions for Achieving Peak Performance. CRC Press, Washington (2005)
3. Petersen, S.A., Divitini, M., Matskin, M.: An Agent-based approach to modeling Virtual Enterprises. International Journal of Production, Planning & Control, Special Issue on Enterprise Modeling 12(3), 224–233 (2001)
4. Rolstadas, A., Andersen, B.: Enterprise Modeling Improvement Global Industrial Competitiveness. Kluwer Academic Publishers (2000)
5. Taveter, K., Wagner, G.: Agent-Oriented Enterprise Modeling Based on Business Rules. In: Kunii, H.S., Jajodia, S., Sølvberg, A. (eds.) ER 2001. LNCS, vol. 2224, pp. 527–540. Springer, Heidelberg (2001)
6. Garlan, D., Schmerl, B.: Model-based Adaptation for Self-Healing Systems. In: Proceedings of the First ACM SIGSOFT Workshop on Self-Healing Systems (WOSS), South Carolina, pp. 27–32 (November 2005)
7. Topol, B., Ogle, D., Pierson, D., Thoensen, J., Sweitzer, J., Chow, M., Hoffmann, M.A., Durham, P., Telford, R., Sheth, S., Studwell, T.: Automating problem determination: A first step toward self-healing computing system. IBM White Paper (October 2003)
8. Hillman, J., Warren, I.: Meta-adaptation in Autonomic Systems. In: Proceedings of the 10th International Workshop on Future Trends in Distributed Computer Systems (FTDCS), Sozhou, China, May 26-28 (2004)
9. Corchado, J.M., De Paz, J.F., Rodríguez, S., Bajo, J.: Model of experts for decision support in the diagnosis of leukemia patients. Artificial Intelligence in Medicine 46(3), 179–200 (2009)
10. De Paz, J.F., Bajo, J., López, V.F., Corchado, J.M.: Biomedic Organizations: An intelligent dynamic architecture for KDD. Information Sciences 224, 49–61 (2013)

11. Rodríguez, S., de Paz, Y., Bajo, J., Corchado, J.M.: Social-based planning model for multiagent systems. Expert Systems with Applications 38(10), 13005–13023 (2011)
12. Bajo, J., De Paz, J.F., Rodríguez, S., González, A.: Multi-agent system to monitor oceanic environments. Integrated Computer-Aided Engineering 17(2), 131–144 (2010)
13. De Paz, J.F., Rodríguez, S., Bajo, J., Corchado, J.M.: Mathematical model for dynamic case-based planning. International Journal of Computer Mathematics 86(10-11), 1719–1730 (2009)
14. Corchado, J.M., Bajo, J., De Paz, J.F., Rodríguez, S.: An execution time neural-CBR guidance assistant. Neurocomputing 72(13), 2743–2753 (2009)
15. Závodská, A., Šramová, V., Aho, A.M.: Knowledge in Value Creation Process for Increasing Competitive Advantage. Advances in Distributed Computing and Artificial Intelligence Journal 1(3), 35–47 (2012)
16. Satoh, I.: Bio-inspired Self-Adaptive Agents in Distributed Systems. Advances in Distributed Computing and Artificial Intelligence Journal 1(2), 49–56 (2012)
17. Agüero, J., Rebollo, M., Carrascosa, C., Julián, V.: MDD-Approach for de-veloping Pervasive Systems based on Service-Oriented Multi-Agent Systems. Advances in Distributed Computing and Artificial Intelligence Journal 1(6), 55–64 (2013)

The Impact of Using Gamification on the Eco-driving Learning

Víctor Corcoba Magaña and Mario Muñoz Organero

Dpto. de Ingeniería Telemática, Universidad Carlos III de Madrid Leganés, Madrid, Spain
vcorcoba@it.uc3m.es

Abstract. This paper analyses and validates the impact of using gamification techniques for improving eco-driving learning. The proposal uses game mechanisms such as the score and achievements systems in order to encourage the driver to drive efficiently. The score is calculated using fuzzy logic techniques that allow us to evaluate the driver in a similar way as a human being would do. We also define the eco-driving tips that are issued while driving in order to help the driver to improve the fuel consumption. Every time the system detects an inefficient action of the driver to a previously known situation such as a bad reaction to a detected traffic sign or a detected traffic accident, it warns the user. The proposal is validated using 14 different drivers performing more than 300 drives with 5 different models of vehicles on 4 different regions of Spain. The conclusions show a positive correlation in the use of gamification techniques and the application of the proposed of eco-driving tips, especially for aggressive drivers. Furthermore, these techniques contribute to avoid drivers coming back to their previous driving habits.

Keywords: Eco-driving learning, Fuel Consumption Optimization, Gamification, Help systems, ITS, Intelligent Vehicle Systems & Telematics, Intelligent systems, User experiments.

1 Introduction

The emission of pollutant gases due to vehicles causes a large number of deaths [1]. On the other hand, the number of old vehicles in circulation have increased exponentially in recent years [2]. Fuel consumption depends on a large number of parameters that can be classified into three major groups: vehicle parameters (engine, aerodynamic and weight), environment (topology, traffic density and weather conditions) and driver (speed, acceleration, deceleration, gear, and air conditioning). Eco-driving is a driving technique based on the setting of the parameters that the user controls. This technique allows us to save fuel regardless of the technology [3] [4].

This paper focuses on two of the most important challenges in eco-driving: motivating the user to drive efficiently avoiding the return to its bad previous driving habits and helping drivers to acquire knowledge about eco-driving in an efficient way. One way to encourage the driver to learn and to apply the eco-driving rules is using

C. Ramos et al. (eds.), *Ambient Intelligence - Software and Applications*,
Advances in Intelligent Systems and Computing 291,
DOI: 10.1007/978-3-319-07596-9_5,

gamification techniques. This method consists of building a game in a non-game context to perform hard and repetitive activities, improving the engagement of user. In [5], we can see a review about gamification and the impact on teaching. There are many scenarios where this concept is applied. For example in [6], authors use emoticons and sounds to encourage the recycling of bottles. When a user throws away a bottle in the garbage, the emoticon smiles. Authors conclude that the proposed system increases the recycling rate by a factor of x3. This concept has also been applied in the field of transport systems. For example, in [7], the authors propose an application to report road accidents.

2 The Eco-driving Game

The proposed solution in this paper consists of an eco-driving assistant in combination with a gamification system. The eco-driving assistant continuously monitors the driver and the environment to propose improvements in his or her driving style in order to save fuel. The eco-driving recommender system is able to use the information coming from different sources (on-board telemetry systems, Internet Web Services and mobile device's embedded sensors such as GPS, accelerometers and camera) to detect the current driving conditions and to adapt the eco-driving tips. On the other hand, when the driver finishes the trip, the system evaluates the quality of the driving in eco-efficiency terms (using fuzzy logic) and assigns a score to the driver. The score can be shared with friends and other users. In addition, we have defined achievements to reward environmental actions and to encourage the continuous use of the eco-driving assistant. The aim of the proposal in this paper is that the user acquires the required knowledge about eco-driving in a user friendly and efficient way, applies the eco-driving tips while driving and does not return to their previous bad driving habits.

The eco-driving assistant runs on a mobile device and which is continuously monitoring the behavior of the driver and the environment. When the eco-driving assistant detects an inefficient action, it notifies the user in order to avoid the same mistake again. In addition, it anticipates upcoming situations on the road in order to avoid the waste of energy if there was a late reaction from the driver. The different recommendations provided by the eco-driving assistant are:

- Driving at a constant speed
- Avoiding sharp accelerations
- Avoiding sharp slowdowns
- Avoiding driving at high revolutions per minute
- Avoiding driving at high speed
- Reducing the intensity in decelerations by anticipating to upcoming situations (such a traffic signal)
- Adapting the speed to environmental factors (rain, wind or important slope angles) or traffic conditions.

Moreover, as noted above, it is essential to encourage and motivate the user to apply the tips and continue using the assistant. We evaluate the driver when it completes the trip from the point of view of energy consumption and we assign a score to it. User scores can be shared with friends and other users establishing a ranking. The gamification techniques are designed based on the goal of obtaining the maximum score. Gamification is the use of game design elements in non-game contexts such as learning environments. The idea is to use concepts from games like: the challenge, the competitiveness and progression in order to motivate the user for improving the driving style from the point of view of energy consumption.

The score of the driver is obtained using a fuzzy logic system. This method allows us to simulate the human knowledge when carrying out certain tasks such as driving. The objective is to model the behavior of an efficient driver. In the model, a set of input variables (acceleration, deceleration, engine speed, standard deviation of vehicle speed, positive kinetic energy and vehicle speed) is involved and the output is the estimation of energy efficiency of a driver. The output variable is a number between 0 and 10. A high value means that the driver is applying the basic rules of eco-driving thoroughly. The proposed system is able to evaluate the driver's driving style based on a knowledge base and the information obtained through the vehicle's diagnostic port (OBD2) [8]. The OBD2 port allows us to obtain the vehicle telemetry. The knowledge base contains the rules that define whether driving is efficient or not. The rules have been obtained through observation of real samples. Our fuzzy system has six rules:

- IF stdSpeed is high AND (acceleration OR deceleration OR speed) is high THEN NonEfficient
- IF engineSpeed is high AND speed is low THEN NonEfficient
- IF acceleration is high and PKI is high THEN NonEfficient
- If engineSpeed is high AND speed is high THEN Efficient
- If stdSpeed is low AND acceleration is low AND PKI is low THEN Efficient
- If stdSpeed is low AND deceleration is low THEN Efficient

On the other hand, we have defined a set of achievements in order to motivate the driver to use the system frequently and in order to allow him to get familiar with ecological challenges. As an example, the user unlocks an achievement when he completes a trip without accelerating sharply. Achievements are a traditional gamification method used to accomplish a certain behavior or to compare the performance of users. Achievements do not normally imply monetary compensation, but they are based on an emotional reward.

3 Evaluation of Eco-driving Game

3.1 Experimental Design

In order to evaluate the improvement in the performance of the eco-driving rules when we use the proposed game, validation tests have been carried out using 5 different vehicle models and 14 driver. Drivers were divided into two groups. The first group did not have the achievement system and the eco-driving assistant only issued eco-driving tips when it detected that the driver was doing inefficient actions from the point of view of energy consumption. The second group made tests with eco-driving game enabled. In this case, when the driver finished the route, the eco-driving assistant assigned a score to the user. The driver could check his score and compare it with the score obtained from friends and other users as well as to see his position in an eco-driving ranking. Tests were made in four different regions of Spain: Madrid, Seville, Granada and "Castile and Leon". However, all tests have performed under similar conditions (road type, number of stops, traffic density and weather conditions) in order to make a fair comparison between drivers.

The eco-driving game was deployed on a Galaxy Nexus mobile device equipped with an ArmV9 processor at 1.2 GHz, 1 GB of RAM and Android 4.1.2. The OBDLink OBD Interface Unit from ScanTool [9].Net was used to get the relevant data (vehicle telemetry) from the internal vehicle's CAN bus. The OBDLink Interface Unit contains the STN1110 chip that provides an acceptable sample frequency for the system. In our tests, we obtain two samples per second. Figure 3 shows an overview of the experimental setting.

3.2 Results

To assess the overall behavior of the driver from the point of view of fuel consumption, we use the fuzzy logic system described above. The score obtained allows us to determine to what degree the user complies with eco-driving rules. A lower score indicates that the driver does not apply the eco-driving rules. In contrast, a high score means that the user is driving efficiently. All drivers performed two tests. The first test is realized before using the eco-driving assistant. After, when the eco-driving assistant has been used 30 times, the second test is made. Table 1 captures the score obtained by drivers without using the eco-driving game feature, before (pre-test) and after (post-test) using the assistant. Table 2 shows the score obtained by drivers who have the game feature enabled on their eco-driving assistants. The score is a number from 0 to 10 where 0 means that the driver is totally inefficient and 10 which is very efficient. We can conclude that when the driver does not have activated the game feature (assistant issues only eco-driving tips), the driving style improves very slightly, and after a short period of time he returns to his previous driving habits. Furthermore, some drivers ignore the eco-driving advice like driver "F1" (aggressive and occasional driver). On the other hand, drivers optimize the driving significantly when the game is enabled and maintain the efficient driving style regardless of the user profile (aggressive, normal, occasional or usual).

In order to validate that the proposal improves the user's motivation to comply with the eco-driving rules and it is not due to random factors, the t-test has been used. Considering the null hypothesis as: "there is no improvement in user score when using the eco-driving game feature" and calculating the p-value we obtain a value of 0.004 (below the 0.05 threshold). Therefore, the null hypothesis (under the 0.05 threshold) can be rejected.

Table 1. User Score without using the eco-driving game

	Driver Profile	Pre-Test	Post-Test	Gain
A1	Aggressive Usual	1	1.5	0.5
B1	Normal Usual	2.1	4.1	2
C1	Normal Usual	2.1	7.8	5.7
D1	Normal Usual	0.9	7.9	7
E1	Normal Usual	1.4	8.7	7.3
F1	Aggressive Occasional	0.1	0	-0.1
G1	Aggressive usual	0.2	2.3	2.1

Table 2. User Score using the eco-driving game

	Driver Profile	Pre-Test	Post-Test	Gain
A2	Aggressive Usual	0,6	8	7,4
B2	Normal Usual	3,7	9,98	6,28
C2	Normal Usual	2,2	10	7,8
D2	Aggressive Occasional	1,3	8,9	7,6
E2	Normal Usual	1,5	9,6	8,1
F2	Normal Usual	2	7,5	5,5
G2	Aggressive Usual	0	8,3	8,3

Table 3 captures the number of drivers who earned each badge. Drivers with the game features enabled, unlocked more achievements than users without the game features active. In addition, they are the only ones who were able to unlock the most complex achievement (unlock all the achievements of the game).

Table 3. Unlocked achievements by drivers

Badget	Game disabled	Game enabled
Obtain 5 points	3 Drivers	7 Drivers
Obtain 7 points	3 Drivers	7 Drivers
Obtain 10 points	0 Drivers	1 Drivers
Complete a lap without decelerating sharply	4 Drivers	5 Drivers
Complete a lap without decelerating sharply over 0.5% of the trip time	6 Drivers	7 Drivers
Complete a lap without accelerating sharply over 1 % of the trip time	2 Drivers	5 Drivers
Complete a lap without accelerating sharply over 2 % of the trip time	6 Drivers	7 Drivers
Complete a lap with a standard deviation less than 2	2 Drivers	6 Drivers
PKI Value over 0.30	0 Drivers	4 Drivers
Average Fuel Consumption equal or less than the value approved by the manufacturer adding 0.2 l/100 Km	1 Drivers	2 Drivers
Unlock all achievements	0 Drivers	1 Drivers

Figure 1 and figure 2 capture the score and the fuel consumption obtained by four drivers while they were using the eco-driving assistant in order to analyse the progression on the learning of eco-driving rules using the proposed method. Two drivers have a normal profile, and the other two have an aggressive profile. These drivers drove in the same route under similar traffic and weather conditions. This route has urban road and highway. Moreover, these tests were performed at 8 A.M. The vehicle was a Citroen Xara Picasso when the game is disabled, and a Ford Fusion 1.4 HDI when the game is enabled.

Figure 1 shows that in the case of drivers with aggressive profile, users slightly improved their driving style when using only the eco-driving advice. However, after an initial improvement, they returned to their previous bad driving habits. On the other hand, when the eco-driving game feature is enabled, aggressive drivers maintain a more optimal driving pattern over the time compared to their initial driving style. Drivers with normal profile also optimized their driving style. The improvement obtained is higher with the game features active.

In figure 2, we can observe that the aggressive driver improves fuel consumption after he has used the eco-driving assistant around 12 times. He even gets, after the initial training, a fuel consumption level similar to that obtained by the normal driver (tests 12-26). However, after this initial improvement, the driver returns to demand the same fuel consumption than in the past. Drivers with normal profile maintained the improvement in fuel consumption over the time.

In the second case (eco-driving game feature enabled), aggressive drivers maintained the improvement in fuel consumption over the time and achieved to save up to 0.9 L/100 Km. Normal drivers decreased up to 0.69 L/100 Km (a lower value than aggressive drivers since they were driving more efficiently and therefore they could not significantly reduce their fuel consumption).

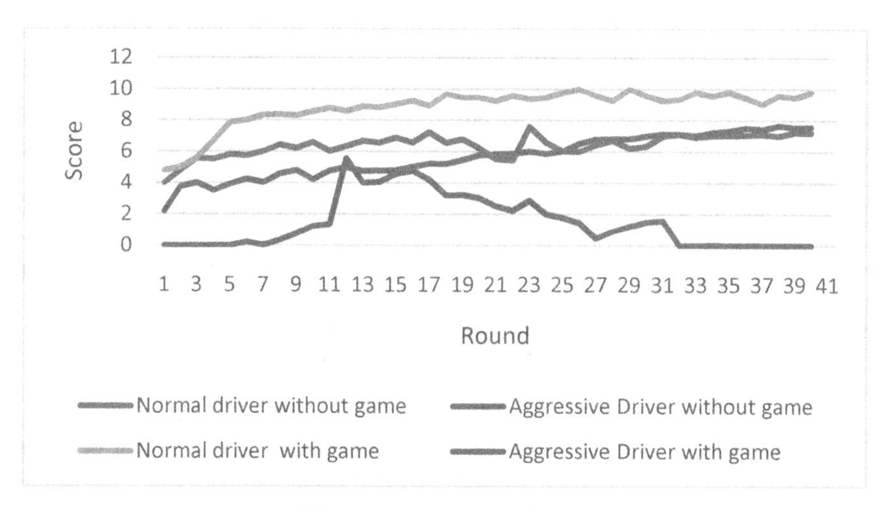

Fig. 1. Evolution of user score

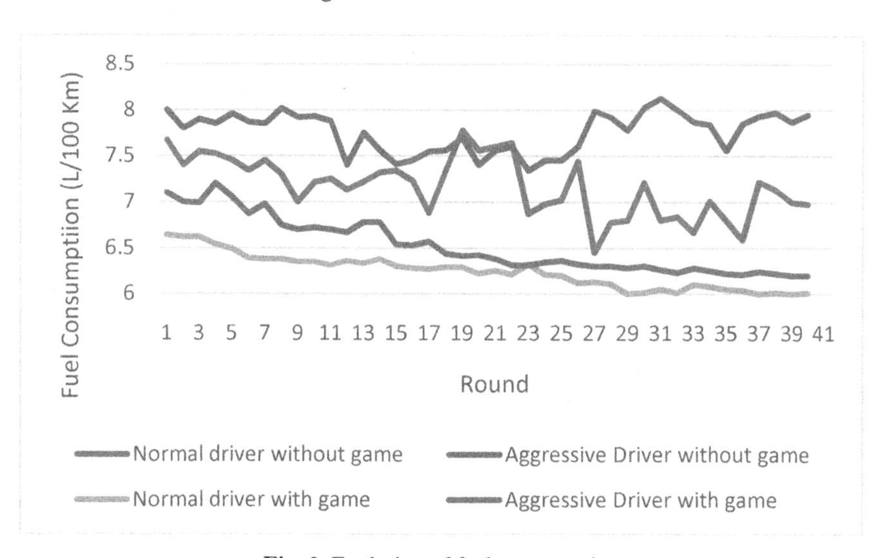

Fig. 2. Evolution of fuel consumption

4 Conclusions

This paper analyses the suitability of using gamification principles to encourage the user to drive efficiently and validates the impacts on fuel consumption obtained when using an eco-driving assistant that, based on the observation of a driver's driving style and comparing it with widely accepted eco-driving rules, proposes recommendations to reduce fuel consumption. The results show that drivers with normal profiles do not need extra motivation to drive efficiently. The eco-driving advice given by the eco-driving assistant are enough to observe an important improvement in their driving style from the point of view of energy efficiency. However, when we use a method to encourage and motivate the drivers such as gamification, drivers tend to improve their driving style even more. Drivers with an aggressive profile fail to improve their fuel consumption and driving style, even if using eco-driving assistants, when not having a motivational reason. During testing, we have seen only a slight improvement during a limited period of time when aggressive drivers use our eco-driving assistant. Moreover, aggressive drivers return to previous bad driving styles after using the assistant. In this case, adding gamification techniques to our eco-driving assistant turned out very useful.

As future work, we want to assess the effects of applying different incentives in the motivation of the driver. The key to the success of the gamification is to identify what incentive is most important for each type of user. For example, on the eco-driving topic, the incentive can be monetary, ecological or safety.

Acknowledgments. The research leading to these results has received funding from the ARTEMISA project TIN2009-14378-C02-02 within the Spanish "Plan Nacional de I+D+I and from the IRENE (PT-2012-1036-370000), COMINN (IPT-2012-0883-430000) and REMEDISS (IPT-2012-0882-430000) projects all funded by the Spanish MINECO Ministry.

References

1. Ericsson, E., Larsson, H., Brundell-Freij, K.: Optimizing route choice for lowest fuel consumption – Potential effects of a new driver support tool. Transportation Research Part C: Emerging Technologies 14(6), 369–383 (2006) ISSN 0968-090X, http://dx.doi.org/10.1016/j.trc.2006.10.001
2. Statistical Office of the European Communities, http://epp.eurostat.ec.europa.eu/portal/page/portal/eurostat/home/ (last access February 2013)
3. Barbé, J., Boy, G.: On-board system design to optimize energy management. In: Proceedings of the European Annual Conference on Human Decision-Making and Manual Control (EAM 2006), Valenciennes, France, September 27-29 (2006)
4. Koskinen, O.H.: Improving vehicle fuel economy and reducing emissions by driving technique. In: Proceedings of the15th ITS World Congress, New York, November 15-20 (2008)
5. Erenli, K.: The impact of gamification: A recommendation of scenarios for education. In: 2012 15th International Conference on Interactive Collaborative Learning (ICL), September 26-28, pp. 1–8 (2012), doi:10.1109/ICL.2012.6402106
6. Berengueres, J., Alsuwairi, F., Zaki, N., Ng, T.: Emo-bin: How to recycle more by using emoticons. In: 2013 8th ACM/IEEE International Conference on Human-Robot Interaction (HRI), March 3-6, pp. 397–397 (2013)
7. Law, F.L., Mohd Kasirun, Z., Gan, C.K.: Gamification towards sustainable mobile application. In: 2011 5th Malaysian Conference in Software Engineering (MySEC), December 13-14, pp. 349–353 (2011), doi:10.1109/MySEC.2011.6140696
8. Godavarty, R., et al.: Interfacing to the on-board diagnostic system. In: 52nd Vehicular Technology Conference, IEEE VTS-Fall VTC 2000, vol. 4, pp. 2000–2004 (2000)
9. OBDLink Adapter, http://www.scantool.net/ (last access June 3, 2014)

Person Localization Using Sensor Information Fusion*

Ricardo Anacleto[1], Lino Figueiredo[1], Ana Almeida[1], and Paulo Novais[2]

[1] GECAD - Knowledge Engineering and Decision Support Research Center, School of Engineering - Polytechnic of Porto, Porto, Portugal
{rmsao,lbf,amn}@isep.ipp.pt

[2] CCTC/Informatics Department, University of Minho, Braga, Portugal
pjon@di.uminho.pt

Abstract. Nowadays the incredible grow of mobile devices market led to the need for location-aware applications. However, sometimes person location is difficult to obtain, since most of these devices only have a GPS (Global Positioning System) chip to retrieve location. In order to suppress this limitation and to provide location everywhere (even where a structured environment doesn't exist) a wearable inertial navigation system is proposed, which is a convenient way to track people in situations where other localization systems fail. The system combines pedestrian dead reckoning with GPS, using widely available, low-cost and low-power hardware components. The system innovation is the information fusion and the use of probabilistic methods to learn persons gait behavior to correct, in real-time, the drift errors given by the sensors.

Keywords: Pedestrian Navigation System, Inertial Navigation System, Indoor Location, GPS, Probabilistic Algorithms.

1 Introduction

A system that is capable of locate an individual can be explored, among others, to improve life quality since emergency teams (fire-fighters, military forces, policeman's and medics) can respond more precisely if the team members location is known, tourists can have more precise recommendations [2], the elderly can be better monitored [13] and parents can be more relaxed with their children.

The motivation for this project emerged from our previous works, where a recommendation system to support a tourist in his vacations has been developed. However, its major limitation is related to obtain tourist location, which is only

* This work is part-funded by ERDF - European Regional Development Fund through the COMPETE Programme (operational programme for competitiveness) and by National Funds through the FCT Fundao para a Cincia e a Tecnologia (Portuguese Foundation for Science and Technology) within project FCOMP-01-0124-FEDER-028980 (PTDC/EEI-SII/1386/2012). Ricardo also acknowledge FCT for the support of his work through the PhD grant (SFRH/DB/70248/2010).

C. Ramos et al. (eds.), *Ambient Intelligence - Software and Applications*, Advances in Intelligent Systems and Computing 291,
DOI: 10.1007/978-3-319-07596-9_6,

based on GPS (Global Positioning System) restricting its use to environments where GPS signal is available [1]. Unfortunately, GPS signal is hardly attenuated by obstacles like walls, canyons composed by high buildings or dense forests.

Therefore a system that allows accurate people location, where GPS signal is unavailable, becomes necessary. There are already some proposed systems that retrieve location in indoor environments. However, most of these solutions require a structured environment. One of the first indoor localization systems was based on electromagnetic sensing [12]. After this many approaches have been developed based on smart floor, RFID, Wi-Fi signal strength, ultrasound and many others. Also, computer science companies, like Google [10] and Microsoft [5], are doing some research on indoor localization systems. Therefore, these systems could be a possible solution for indoor environment, but in a dense forest or urban canyons they are very difficult to implement.

To suppress structured environment limitations, an Inertial Navigation Systems (INS) can be developed. An INS is constituted by accelerometers, gyroscopes and other type of sensors based on MEMS (Microelectromechanical systems), which are tiny and lightweight making them ideal to integrate in the person's body. These systems are based on the Pedestrian Dead Reckoning (PDR) technique and the sensors are spread along the person's body to gather acceleration and direction values to estimate the person's walking path. Unfortunately, large deviations of inertial sensors can affect performance, so the INS systems big challenge is to correct the sensors deviations. A module working only with PDR is not able to ensure that the geographical positions are accurate within a few meters.

To reduce these typical errors, Feliz et al. [7] estimates the error in each step of a pedestrian walks to ensure that the small error produced, when speed and position are estimated, will not influence the speed and position estimation for the next step. Castaneda and Lamy-Perbal [4] proposes a fuzzy logic procedure for better foot stance phase detection. Bebek et al. [3] introduce a high-resolution thin flexible ground reaction sensor, which measures zero velocity duration to reset the accumulated errors.

Despite all of these systems can provide localization everywhere there are still lack of location accuracy and some improvements can be made. This paper presents our proposal that includes force sensors and learning algorithms that gathers previous knowledge of the person steps to "self-learn" the user gait behavior, *e.g.*, using GPS when it has very good signal and low error rate to calibrate/learn the INS system. More detailed information about the system architecture will be presented in section 2. Section 3 presents some experimental results and in section 4 are presented some conclusions and the future work.

2 System Architecture

The main claim of our proposal is the capability to retrieve location everywhere, independently of the environment and only based in sensors that are placed in the human body.

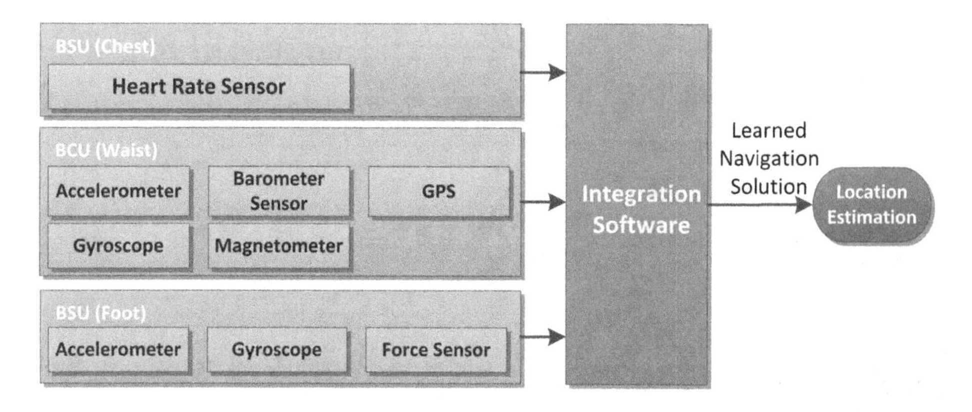

Fig. 1. System Software Architecture

Our system is constituted by two parts (figure 1), which will be discussed on the next sub-sections:

- Hardware - consists on Body Sensors Units (BSU) placed on the person foot, waist and chest to collect movement's data. These BSU's communicate with a Body Central Unit (BCU) via a wireless network (section 2.1);
- Software - is composed by two parts, the sensor fusion incorporated on the BCU (section 2.1), which integrates the information from sensors and thereby tries to estimate the person location. The other part, implemented on the mobile device (section 2.2), is constituted by the learning algorithms.

Two quantifiable success criteria were defined to this project, the first one is the accuracy that must be between 90% and 95%, and the second one is the delay between the sensor readings and the exhibition of the current user location. To be considered real-time this delay should be less than 2 seconds.

2.1 Body Sensor Units and Sensor Fusion

Small BSU are placed on the person body to collect information about body movements. In the future we want to integrate these sensors into person's clothes and shoes, to be more imperceptible to the user. This data is sent, through a wireless network based on ZigBee [9], to the BCU that handle the calculations to estimate, in real-time, the person location. The sensors were developed based on the Smart Sensors philosophy, connected to an integrated circuit module to pre-process and codify the collected signal. The BCU module also sends data from the INS system to a mobile device via a Bluetooth connection.

The BSU's are distributed like this, one in the foot that include a force sensor, a gyroscope and accelerometer (figure 2), one in the chest area that include a heart rate sensor, and another in the abdominal/waist area that contains an accelerometer, a gyroscope and a barometer. Force sensors were included since they can improve the detection of the moment when the user touches his feet on

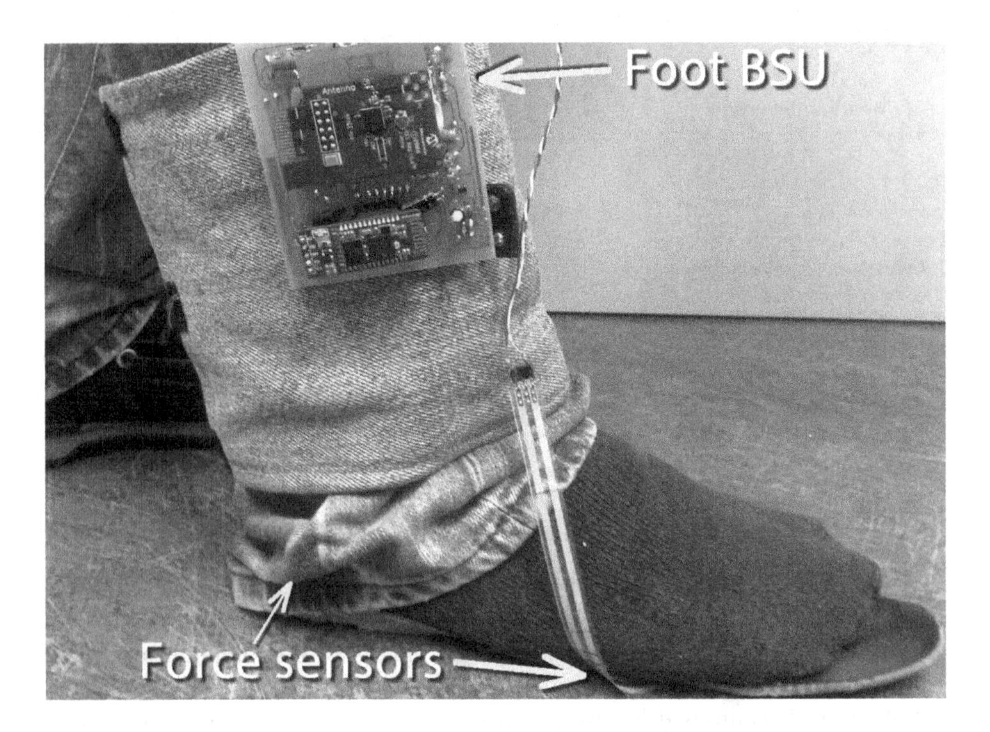

Fig. 2. Foot Body Sensor Unit (BSU)

the ground, as well as, the correspondent contact force, which combined with the accelerometer (used to obtain the step acceleration) provide a more exact step length calculation. The gyroscope is valuable to get the body travel direction, as well as, to transform the acceleration data to the navigation frame. The barometer is used to obtain user elevation. A heart rate sensor is used to know more precisely the user activity (*e.g.*, walking, running, etc.), thus improving the system precision, since when a person is traveling faster the heartbeat is higher than when a person is only standing-up.

In order to have a successful implementation of this wireless network and the corresponding sensors there are some "open problems" that still must be solved. These problems include issues related to deployment, security, calibration, failure detection and power management.

Having several sources of data can be useful to detect more accurately all the person movements, however this integration can be very difficult to implement. From our experience with the INPERLYS project [8], which uses a MEMS accelerometer to estimate the traveled distance and a digital compass for orientation, an INS algorithm working by itself doesn't provide a good location accuracy due to sensors drift errors.

In our approach there are three important software pieces, a preprocessing algorithm on the sensors to remove some noise, a Kalman filter based algorithm to fuse the data from the sensors and a "Learning Algorithm".

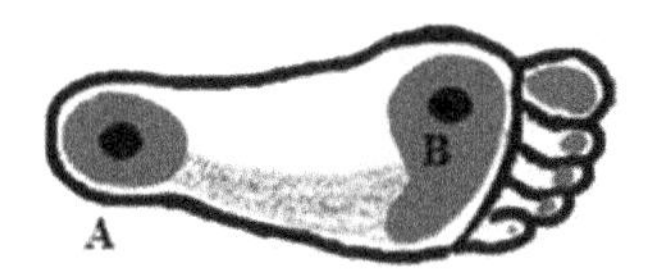

Fig. 3. Force sensors location on the foot

Typically the drift is mitigated taking advantage of the Zero Velocity Update (ZUPT) [11], meaning that the integration of inertial measurement is only performed during the swing of legs and the velocity errors can be reset at each step since when INS is stationary the true velocity must be zero. This technique is used by several systems and typically an accelerometer is used to detect when the foot touches the ground. However, our system uses the force sensors to detect it, as well as, the respective contact force. This force information can be also useful to improve the detection of the person activity type (running, walking fast, etc.).

From figure 3 the force sensors position on person foot can be visualized. These positions were chosen since they provide information about when the foot touches the ground (A) and when foot leaves the ground (B). This corresponds to the stance phase of the gait cycle and is during this phase that zero velocity occurs, so the acceleration data isn't considered during this period. These are also the zones where more force is applied on the foot.

2.2 Learning System and Mobile Application

The learning algorithms and a localization module are implemented at the mobile device. In order to reduce errors provided by MEMS, a learning algorithm is being developed to learn the walking/moving behaviors in each type of environment for correcting, in real-time, the data gathered from the sensors.

Walking is a cyclic activity, which represents a cyclic pattern of movement that is repeated over and over, step after step [14] [16]. So, these walking patterns can be extracted in the learning phase and used as a reference model. These patterns are learned, over the time, when GPS is available (with very good signal) or in a controlled phase where from a set of exercises the gait analysis is obtained. Resuming, the system will be always learning the step pattern and will be improving it over the time.

Besides the learning module, another module was developed (for Android OS) to integrate the data from the different sources, and is responsible to retrieve the current user's location estimated by the GPS/INS localization system.

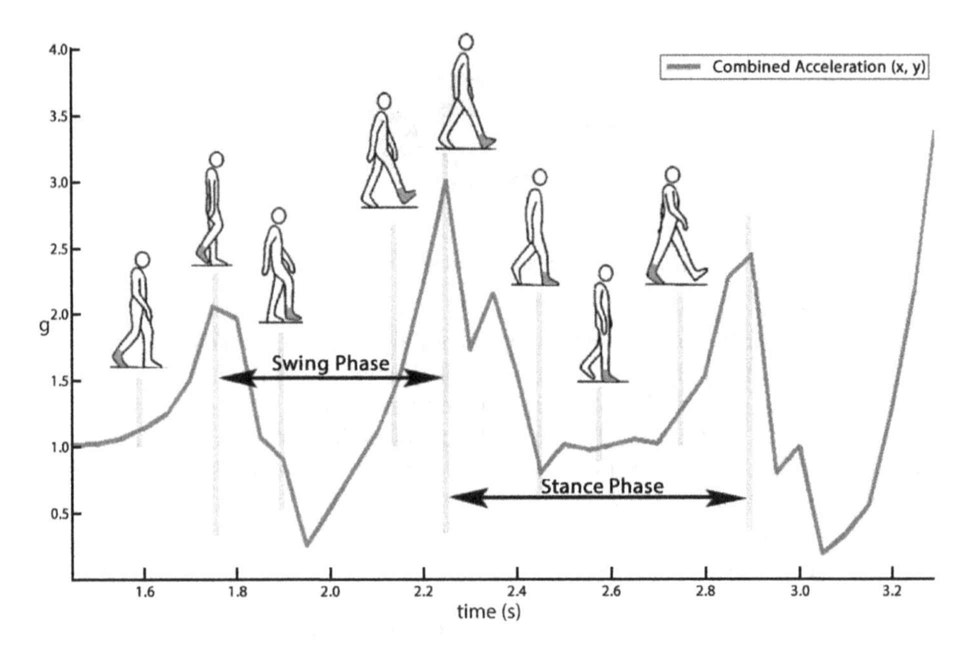

Fig. 4. Foot acceleration during the gait cycle

3 Experimental Results

Since the start of this project the research team has been studding and experimenting some implementations, techniques and technologies on pedestrian navigation. With this acquired knowledge and with the encountered problems, we found that an INS solution only based on accelerometers and gyroscopes is not accurate enough.

First of all, the sensors and their positioning. As stated earlier, an INS can bring several problems, especially because of the sensors drift. Also, because of people different sizes, the ideal sensors position to one person can be different to another. This leaves to another challenge, the discovery of an ideal spot for each sensor to work in diverse types of persons. We have found, according to our tests and results, that the ankle is a very good position to put the foot BSU [15], as can be seen on figure 2.

According to the literature the gyroscope bias is the cause of the most of the horizontal errors, so a good practice is to recalibrate the gyroscopes before each experiment. Another sensor that has significant errors is the accelerometer. Figure 4 presents the results of our experiments to determine the acceleration pattern during a gait cycle in a normal walking. Faulkner et al. [6] also have observed that the accelerometer can register large signal peaks. This peaks can introduce some errors to the system if the accelerometer doesn't provide a good acceleration range.

This range can reach $\pm 10g$ when walking and $\pm 13g$ when running or climbing stairs. For best results, accelerometers with at least a $\pm 10g$ range and gyroscopes with a $\pm 900\,^{\circ}/s$ range should be used, which nowadays are a relatively high specifications for MEMS.

A lot of works suggests that the stance phase detection can be tricky, since the accelerometer takes some time in order to obtain that information or due to sensor deviations. In our system force sensors were used to try to overcome this problem, which proved to be a good approach. This sensors are mainly used to distinguish the swing and the stance phase from the gait cycle. It is during the stance phase that zero velocity must occur, since during this phase the foot doesn't move, so no displacement must be considered.

Our tests began with a first approach that uses the Kalman filter in conjunction with a ZUPT module working only with the gyroscope and accelerometer data. These tests were performed on a room involving a total traveled distance of 50 meters. This first approach demonstrated an error of 9.2% on a normal walk path.

After this test the ZUPT module was implemented to only work with the force sensor data. With the use of force sensors the stance phase was better detected, which helped to improve the system errors by 1.3%, this means that the estimated error have reduced to 7.9%.

4 Conclusion

Develop an accurate, inexpensive, small and unobtrusive localization system to be used by persons, when they are on foot, in environments where GPS is unavailable can be a huge challenge. Many approaches already have been proposed, but must of them rely on a structured environment that usually is unfeasible to implement and the other's don't provide the necessary accuracy.

In this work it was used a set of small MEMS sensors and the available data was explored to the maximum in order to provide an acceptable level of performance. In the described solution these sensors were spread along the body to detect the person movements in two places, foot and waist.

Since the detection of stance phase using accelerometers and gyroscopes can introduce several errors on INS, our proposal includes force sensors on foot plant to improve the stance detection, and so improve system accuracy. The results from our first experiments, which involved a walk of 50 meters, are very satisfactory, the use of force sensors allowed an average error reduction of 1.3%.

However, a learning algorithm is being developed to improve, even more, the overall system accuracy. This algorithm will learn the person gait cycle when GPS is available or in a learn phase walk, to then in real-time perform corrections in the INS, thus improving INS accuracy.

Another important point of an INS is to inform, in real-time, the user's current location and not only record positions for future walking path analysis. However, sometimes this task is difficult to implement due to the process complexity mainly because of sensor data acquisition delays, communication delays and data processing execution that can take some time. This introduces a significant delay between the real and the processed location (that appears to the user on the mobile device).

References

1. Anacleto, R., Figueiredo, L., Luz, N., Almeida, A., Novais, P.: Recommendation and planning through mobile devices in tourism context. In: Novais, P., Preuveneers, D., Corchado, J. (eds.) Ambient Intelligence - Software and Applications. AISC, vol. 92, pp. 133–140. Springer, Heidelberg (2011)
2. Anacleto, R., Luz, N., Figueiredo, L.: Personalized sightseeing tours support using mobile devices. In: Forbrig, P., Paternó, F., Mark Pejtersen, A. (eds.) HCIS 2010. IFIP AICT, vol. 332, pp. 301–304. Springer, Heidelberg (2010)
3. Bebek, O., Suster, M.A., Rajgopal, S., Fu, M.J., Huang, X., Cavusoglu, M.C., Young, D.J., Mehregany, M., van den Bogert, A.J., Mastrangelo, C.H.: Personal navigation via high-resolution gait-corrected inertial measurement units. IEEE Transactions on Instrumentation and Measurement 59(11), 3018–3027 (2010)
4. Castaneda, N., Lamy-Perbal, S.: An improved shoe-mounted inertial navigation system. In: 2010 International Conference on Indoor Positioning and Indoor Navigation (IPIN), pp. 1–6 (2010)
5. Chintalapudi, K., Padmanabha Iyer, A., Padmanabhan, V.: Indoor localization without the pain. In: Proceedings of the Sixteenth Annual International Conference on Mobile Computing and Networking, MobiCom 2010, pp. 173–184. ACM, New York (2010)
6. Faulkner, W., Alwood, R., Taylor, D., Bohlin, J.: Altitude accuracy while tracking pedestrians using a boot-mounted IMU. In: IEEE/ION Position Location and Navigation Symposium (PLANS), pp. 90–96 (May 2010)
7. Feliz, R., Zalama, E., Gmez, J.: Pedestrian tracking using inertial sensors. Journal of Physical Agent 1, 35–42 (2009)
8. Ferreira, H., Figueiredo, L.: INPERLYS - independent personal location system. In: Cech, P., Bures, V., Nerudova, L. (eds.) Ambient Intelligence and Smart Environments, Ambient Intelligence Perspectives II, vol. 5, pp. 93–100 (2009)
9. Kinney, P.: Zigbee technology: Wireless control that simply works. In: Communications Design Conference, vol. 2 (2003)
10. McClendon, B.: A new frontier for google maps: mapping the indoors (2011)
11. Nilsson, J., Skog, I., Handel, P.: Performance characterisation of foot-mounted ZUPT-aided INSs and other related systems, pp. 1–7. IEEE (2010)
12. Raab, F., Blood, E., Steiner, T., Jones, H.: Magnetic position and orientation tracking system. IEEE Transactions on Aerospace and Electronic Systems (5), 709–718 (1979)
13. Ramos, J., Anacleto, R., Costa, Â., Novais, P., Figueiredo, L., Almeida, A.: Orientation system for people with cognitive disabilities. In: Novais, P., Hallenborg, K., Tapia, D.I., Rodríguez, J.M.C. (eds.) Ambient Intelligence - Software and Applications. AISC, vol. 153, pp. 43–50. Springer, Heidelberg (2012)

14. Saunders, J., Inman, V., Eberhart, H.: The major determinants in normal and pathological gait. The Journal of Bone & Joint Surgery 35(3), 543–558 (1953)
15. Terra, R., Figueiredo, L., Barbosa, R., Anacleto, R.: Step count algorithm adapted to indoor localization. In: Proceedings of the International C* Conference on Computer Science and Software Engineering, C3S2E 2013, pp. 128–129. ACM, New York (2013)
16. Vaughan, C., Davis, B., O'connor, J.: Dynamics of human gait. Human Kinetics Publishers Champaign, Illinois (1992)

Improving Modularity, Interoperability and Extensibility in Ambient Intelligence

Marco Gomes, Davide Carneiro, André Pimenta, Milton Nunes,
Paulo Novais, and José Neves

CCTC/DI - Universidade do Minho
Braga, Portugal
{marcogomes,dcarneiro,apimenta}@di.uminho.pt,
pg22797@alunos.uminho.pt, {pjon,jneves}@di.uminho.pt

Abstract. Ambient Intelligence (AmI) and its related fields emerged some years ago with the exciting promise of pervasive intelligence, magic interaction mechanisms, and everywhere availability. This promise would be materialized in homes that knew all about our habits and preferences, proactive workplaces to support people's work or personal digital assistants to improve our daily living in all aspects possible. This somewhat utopian vision, expected by many to have already taken place, remains unaccomplished and far from it. Many challenges still lay ahead which delayed and continue to delay the expected technological unravelling. In this paper we focus on the immense technological challenges of designing and implementing AmI Systems. Specifically, we propose a technological approach that will contribute to overcome some of these challenges by making developed AmI solutions more modular, interoperable, and extensible. This will result especially advantageous for large development teams or teams that span multiple institutions.

Keywords: Ambient Intelligence, Interoperability, Switchyard.

1 Introduction

Ambient Intelligence is one of those sub-fields of Artificial Intelligence that stimulates our creativity. It results very easy for us to imagine scenarios in which the artefacts around us have intelligence or consciousness, constantly interact with us in a natural way and are always available. Some of these examples have moved from the imagination of book writers and movie producers to the paper or screen, to result in pieces describing a possible and very appealing future, one of the most popular examples being the futuristic world depicted in the film *Minority Report*. Here, Captain John Anderton interacts with a series of futuristic interfaces and intelligent tools to assist in his fight against (still to-be-committed) crimes. In the books, examples of an exciting future can be found, for example, in the fictional universe of *The Hitchhiker's Guide to the Galaxy*. In this world doors, for instance, are conscious (although their single ability is to feel valued when people use them).

C. Ramos et al. (eds.), *Ambient Intelligence - Software and Applications*, Advances in Intelligent Systems and Computing 291,
DOI: 10.1007/978-3-319-07596-9_7,

While the second example is extracted from a humorous piece, the first allows to think quite seriously on the future that awaits us. Hopefully. Indeed, the actual broad implementation of even the simplest examples seen in these pieces seems still distant in time. Moving from fiction to more serious grounds, the same issue exists: the envisioned new world fostered by Ambient Intelligence and "foreseen" by Bogdanowicz et al. [7] is still far from reality.

Indeed, many of the technological requirements and challenges pointed out by the authors still remain nowadays or are only partly solved. Meanwhile, as depicted in the following section, new challenges emerge that need to be addressed for the sake of the reliability and acceptance of such systems. This paper makes an analysis of these challenges, with a particular focus on technological challenges. We propose an approach based on the novel SwitchYard framework to facilitate the development of more modular and extensible AmI systems. The main aim is to empower development efforts by distributed teams and the technology-independent integration of different systems or modules, to foster the development of AmI.

2 Current Challenges in AmI

As stated in the introductory section, there are several challenges that are, still today, holding AmI development back. One of these challenges, often disregarded by computer scientists (who form the backbone of AmI development) concerns privacy, identity and security issues. In [4] the authors make a thorough analysis of 70 AmI projects, principally in Europe, concerning these issues. They conclude that in general, current projects present a rather too sunny view of our technological future, ignoring or postponing dealing with some pressing issues. The authors also make an interesting reference to the SWAMI project (Safeguards in a World of Ambient Intelligence) which, against this trend, has constructed what they deemed "dark" scenarios [8], to show how things can go wrong in AmI and where safeguards are needed. As Rouvroy puts it, the challenge here is to preserve the individual freedom to build one's own personality without excessive constrains and influences while have control over the aspects of one's identity that one projects on the world [6].

Marzano, on a different view, looks at the cultural implications of an unregulated or indiscriminate growth of AmI, making a parallel with the industrial revolution [1]. As, at the time, more was (later proved to be) not necessarily better (take for instance consequences such as the pollution), right now, *smarter* may also not be necessarily better. Indeed, we may simply not want a smart juicer or a talking toaster.

But let us focus on the technological challenges that are still ahead. One the one hand, we have the challenges that are related with the physical constraints and nature of the necessary hardware. In [3] the authors examine the intricate relationship between the growing need for more computational and communicational power to support increasingly complex services and, at the same time, the need for smaller, more lightweight and efficient devices. It is easy to understand how the objectives of these two fields conflict.

Another issue holding back a faster development of AmI is the scatter of research efforts. Indeed there are currently many different institutions doing research on very similar topics. When these institutions want to conciliate efforts they may find it difficult to do so since they use different technologies, standards or approaches. We believe that facilitating this integration and interoperability could result in a coming closer of different teams, whom could join efforts and more efficiently work together for the same goal. With this objective in mind, we are developing an open architecture to support an AmI system: open not only in the sense that it relies on open software but also, and most importantly, that it can easily integrate external services, as well as provide its own to external requesters. This architecture and its main advantages are described in the following sections.

3 Architecture

In our pursuit to develop an architecture that covers the main AmI technological needs, we first definite it at a conceptual level. The proposed architecture is logically divided into several packages that encapsulate a set of features and tasks. Figure 1 presents its high-level view, detailing the five packages that compose the system.

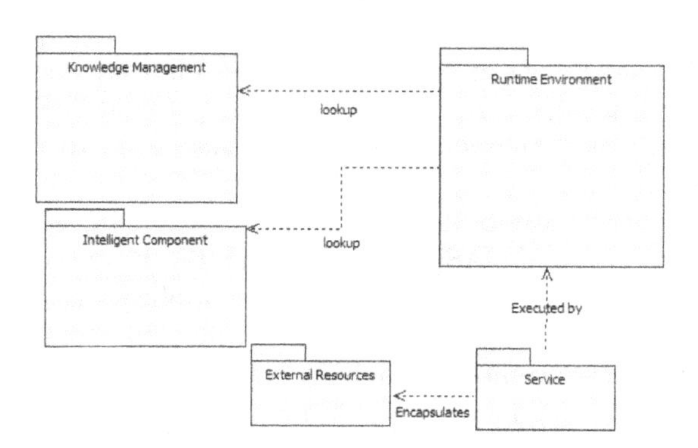

Fig. 1. High-level view of the architecture

The Runtime Environment is the main component of the architecture, where system operations are executed. It is also through this environment that the remaining components are accessed. The Service component, executed by the Runtime Environment, contains all services of monitoring and data collection through the use of sensors, and is responsible for encapsulating all external resources collected by the system, represented by the External Resources. The tasks carried out by the Runtime Environment are also supported by queries to components Knowledge Management and Intelligent Component. Knowledge Management supports the management of information and knowledge collected by the system. The Intelligent Component contains components of Intelligent Systems/Artificial Intelligence used in this architecture for data processing.

The Runtime Environment corresponds to the core component of the architecture where the main features are executed. It is also divided in two subcomponents, which are nonetheless interconnected: the Runtime Management and Runtime Platform. The first is a platform for managing processes running on the system. It is composed of a Data Manager that contains an interface to link the Knowledge Manager, an Execution Manager responsible exclusively for the management of the execution of system processes and the Service Directory that contains all information from the execution of services in the architecture. It should also be noted that the Runtime Management has an interface to connect the Service component. The Runtime Platform is a platform for the execution of the Runtime Environment. This is composed of a Service Bus that, through the interface provided by the Runtime Management, establishes communication between the two subcomponents and also a Service Execution Engine that represents the execution engine services in the architecture.

AmI systems are commonly described as electronic environments that seamlessly interact and adapt to human needs, in which people are surrounded by intelligent and intuitive interfaces embedded in all kinds of objects. To take full advantage of the information gathered ubiquitously from various sources in the environment there is a need for a software infrastructure that allows an easy integration, promotes interoperability, and focuses on extensibility. Considering these aspects, in this work we present an infrastructure to support an efficient approach for the development of AmI applications, following an approach based on Service Oriented Architectures (SOAs). Indeed, SOAs are being increasingly adopted in both the academic and industrial arenas, even to integrate Multi-agent Systems [2].

The more appropriate way of doing so is to adopt a Service-Component Architecture (SCA): a group of OASIS specifications that has become an industry standard. It is intended for the development of applications based on SOA, which defines how computing entities interact to perform work for each other. Originally published in November 2005, SCA is based on the notion that all the functions in an system should exist in the form of services that are combined into composites to address specific business requirements. In other words, it allows to build service-oriented applications as networks of service components. SCA is used for building service components, assemble components into applications, deploy to (distributed) runtime environments and reuse service components built from new or existing code using SOA principles. This approach is advantageous in AmI for the following reasons:

- Interoperable. Provides loose coupling allowing to integrate without need to know how components are implemented. Components can be written using any language, and can use any communication protocols and infrastructure to link them, making it easier to integrate components to form composite applications.

- Maintainable. Composition of solutions is clearly described as declarative application of infrastructure services. Simplification for all developers, integrators and application deployers.
- Flexibility of Development. Service Components are easier to develop because the semantics of each independent Service Component are significantly less complex than the overall of a single, (relatively large) monolithic application; each Service Component can be developed by a different team of developers, each of whom focus only on their component without having to know the details of work done by others. Components can easily be replaced by other components and services can be easily invoked either synchronously or asynchronously.
- Reuse. Since each Service Component has well-defined interfaces, each component can be developed, tested and debugged independently of the other components. This not only speeds up project implementations but, in the case of well-designed Service Components, also leads to significantly enhanced reuse.
- Dynamic Deployment and Runtime Modification/Replacement. Service Components can be dynamically deployed to remote nodes at runtime, and components within a process can be easily replaced by new or updated components, further reducing the time taken to modify or change an existing process in response to business requirements.
- Configuration Management and Version Control. Service Components facilitate version control and dynamic configuration management, allowing fine-grained control over deployments across the enterprise.

SCA provides a good basis for AmI applications [9], it is in line with our architectural model and it fulfils major AmI deployment requirements by promoting late bindings at deploy time and runtime with the support of several relevant technologies including POJO, SOAP, REST, BPMN, BPEL, JMS, Camel or Rules services. But most of all it is currently supported by several major commercial and open source products such as Jboss Switchyard, IBM WebSphere or TRENTINO (C++).

From the several available implementations of SCA we have chosen JBoss SwitchYard since it is an open source solution in a relative mature state, and also enhances some of the SCA advantages. Specifically, Switchyard advocates transparency when running a service during its whole lifecycle. Important aspects such as connectivity, orchestration and routing do exist on SwitchYard in a modular format, which means one can deploy them in an independent way. Using a SwitchYard graphical user interface (Fig. 2), one can build visual models of the applications, that are meant to improve the software engineer's ability to comprehend and communicate the full composition of their applications and also to speed up development and integration projects.

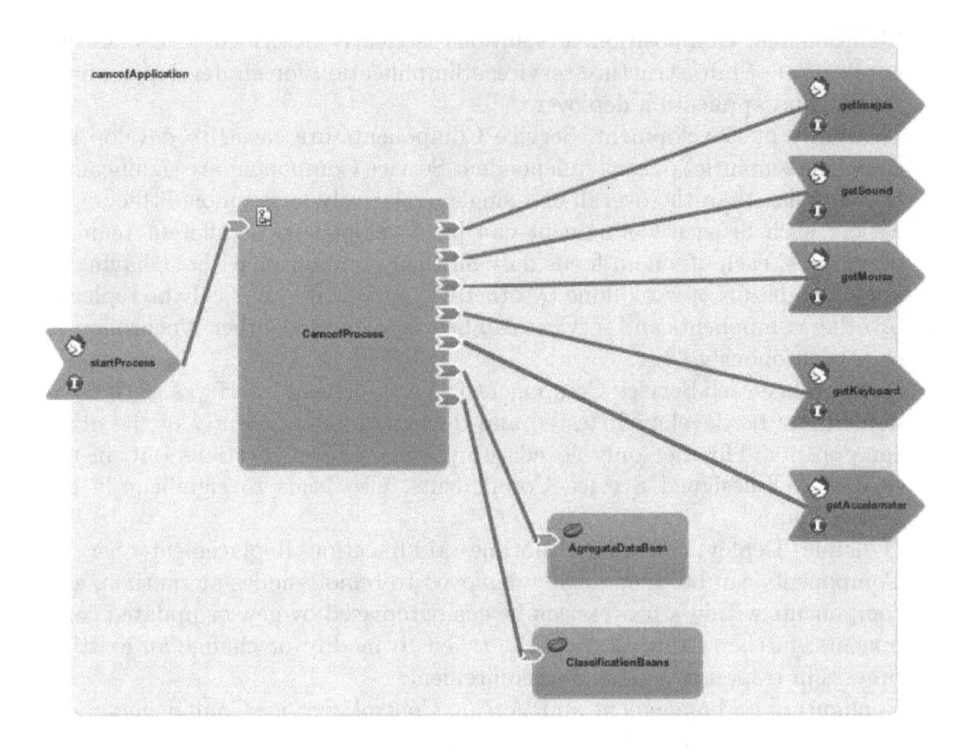

Fig. 2. Application composition using SwitchYard graphical user interface

4 Case-Study of Context-Aware Multimodal Communication system

This case study describes an application in which contextual information about the user is collected and used to detect states of stress and fatigue. The purpose is to enrich communication processes allowing for its users to communicate in ways that are closer to face-to-face communication. The estimation of stress and fatigue are based on the transparent analysis of the user's behaviour and interaction patterns [5]. In gathering data the following sensors were involved:

- **Accelerometer** - These devices, placed on the chair, keyboard and mouse, measure how the user is moving and the amount of force he is applying in the peripherals;
- **Mouse and Keyboard** - These devices provide information about how the user interacts with the peripherals (e.g. velocity of the mouse, typing rhythm, number of mistakes).
- **Microphone** - Microphones are used to measure the amount of noise in the vicinity of the user, allowing to perceive their social environment.
- **Video Camera** - An estimation of the amount of movement is calculated from the video camera. The image processing is based on difference techniques to calculate the amount of movement between two consecutive frames.

In order to materialize the architecture we developed a concrete instance for this specific application. All the described sensors are encapsulated by services running locally. Each of these services exposes different features using a Web Service interface. These services were integrated using SwitchYard allowing the service orchestration to support the automation of system processes by loosely coupling services across different applications. There is a clear separation between process logic and Web Services, providing the system with increased flexibility.

The main process was modelled using Business Process Modelling Notation (BPMN). In this standard (Fig. 3), a consumer service invokes a process flow via the service interface. The orchestration engine invokes services to process various service which in turn invoke further service requests until the workflow process is completed and results are provided to the service consumer. The service orchestration engine, a component of the SwitchYard, handles the overall process flows, calling the appropriate web services and determining the next steps to complete.

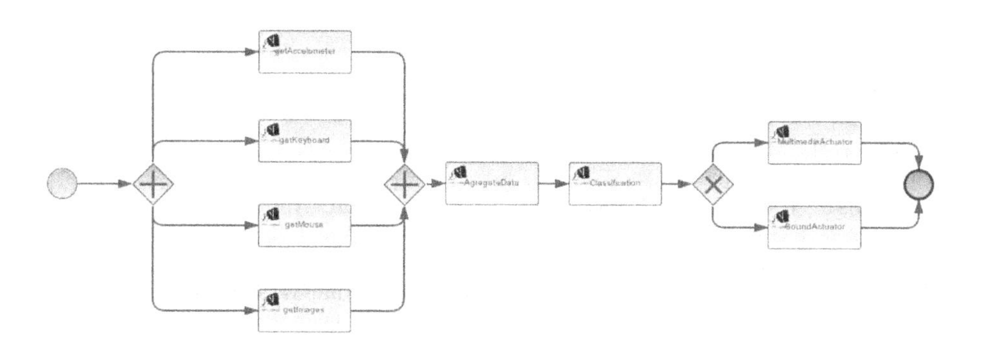

Fig. 3. The "camcof" process, view through BPMN

With this architecture it is possible to orchestrate the acquisition, transformation, and classification processes used to collect and to extract meaningful information. It is also possible to use different services (encapsulating different and disperse sensors), simultaneously or individually, coordinating all the elements of the main process.

5 Conclusions

Ambient Intelligence hasn't developed with the expected pace due to a number of challenges that were pointed out at the beginning of this paper. Many of these challenges still exist nowadays and include the difficulty in integrating and incorporating different approaches as well as the difficulty in developing highly modular systems that can be easily configurable and extended at need.

In this paper we presented an approach targeted at such scenarios. One that allows to build AmI systems in a decoupled way, without technological constraints, relying on a service-oriented approach. This approach is also distributed in the sense that services can be running anywhere in the world. For the manager, this is completely transparent. Nonetheless, high-level functionalities can be built when these services are combined using appropriate rules.

We believe that approaches such as this will not only make it easier to build larger-scale AmI systems but also increase the opportunity for researchers to more easily integrate their work, leading to more complex and AmI systems, with far richer functionalities. Moreover, services scattered around the world can transparently be used to achieve this goal in a far easier way.

Acknowledgements. This work is part-funded by ERDF - European Regional Development Fund through the COMPETE Programme (operational programme for competitiveness) and by National Funds through the FCT - Fundação para a Ciência e a Tecnologia (Portuguese Foundation for Science and Technology) within project FCOMP-01-0124-FEDER-028980 (PTDC/EEI-SII/1386/2012).

References

1. Aarts, E.H., Marzano, S. (eds.): The new everyday: Views on ambient intelligence. 010 Publishers (2003)
2. Aguero, J., Rebollo, M., Carrascosa, C., Julin, V.: MDD-Approach for developing Pervasive Systems based on Service-Oriented Multi-Agent Systems. Advances in Intelligent Computing and Artificial Intelligence Journal 1(6) (2013)
3. De Man, H.: Ambient intelligence: gigascale dreams and nanoscale realities. In: IEEE International Solid-State Circuits Conference, ISSCC. Digest of Technical Papers, pp. 29–35. IEEE (2005)
4. Friedewald, M., Vildjiounaite, E., Punie, Y., Wright, D.: The Brave New World of Ambient Intelligence: An Analysis of Scenarios Regarding Privacy, Identity and Security Issues. In: Clark, J.A., Paige, R.F., Polack, F.A.C., Brooke, P.J. (eds.) SPC 2006. LNCS, vol. 3934, pp. 119–133. Springer, Heidelberg (2006)
5. Pimenta, A., Carneiro, D., Novais, P., Neves, J.: Monitoring Mental Fatigue through the Analysis of Keyboard and Mouse Interaction Patterns. In: Pan, J.-S., Polycarpou, M.M., Woźniak, M., de Carvalho, A.C.P.L.F., Quintián, H., Corchado, E. (eds.) HAIS 2013. LNCS, vol. 8073, pp. 222–231. Springer, Heidelberg (2013)
6. Rouvroy, A.: Privacy, data protection, and the unprecedented challenges of ambient intelligence. Studies in Ethics, Law, and Technology 2(1) (2008)
7. Bogdanowicz, M., Scapolo, F., Leijten, J., Burgelman, J.C.: Scenarios for ambient intelligence in 2010, pp. 3–8. Office for official publications of the European Communities (2001)
8. Wright, D.: The dark side of ambient intelligence. Info 7(6), 33–51 (2005)
9. Giner, P., Pelechano, V.: An Architecture to Automate Ambient Business System Development. In: Aarts, E., Crowley, J.L., de Ruyter, B., Gerhäuser, H., Pflaum, A., Schmidt, J., Wichert, R. (eds.) AmI 2008. LNCS, vol. 5355, pp. 240–257. Springer, Heidelberg (2008)

UAVs Applied to the Counting and Monitoring of Animals

Pablo Chamoso, William Raveane, Victor Parra, and Angélica González

Computer and Automation Department, University of Salamanca, Spain
{chamoso,wr,parra,angelica}@usal.es

Abstract. The advantages of intelligent approaches such as the conjunction of artificial vision and the use of Unmanned Aerial Vehicles (UAVs) have been recently emerging. This paper presents a focused on obtaining scans of large areas of livestock system. Counting and monitoring of animal species can be performed with video recordings taken from UAVs. Moreover the system keeps track of the number of animals detected by analyzing the images taken with the UAVs cameras. Several tests have been performed to evaluate this system and preliminary results and the conclusions are presented in this paper.

Keywords: Unmanned Aerial Vehicle, Convolutional Neural Networks, livestock detection.

1 Introduction

Possibilities of applying UAVs in the professional world have increased in recent years thanks to advances in technology, especially aerial imaging. Aerial imaging provides detailed images and quick monitoring of large areas, making it a very efficient way to solve the problem proposed in this study.

The problem presented in this study is related to cynegetic activities. So far, the different methods that are carried out for conducting censuses to optimize hunting and harvesting in farms and private reserves consists of conducting pathways, which rely on feeders and hunting results.

The case study presented in this paper adapts the developed system as a methodology for conducting population censuses of farm animals. A farm with cows was used for this case study. Area sweeping techniques in combination with visual detection, recognition techniques and UAV system control are used and will be explained throughout this article.

An important part of the system is its ability to visually identify targets on the ground. As with any computer vision application, this is a difficult problem to solve due to the high dimensionality of the input data. The numerical representation of a photograph will vary greatly with any subtle change in illumination, camera conditions, or even object appearance, among other factors. Shadows cast by the target objects only complicate matters as they usually impede effective segregation from the terrain background. These conditions can be overwhelming for most traditional image

C. Ramos et al. (eds.), *Ambient Intelligence - Software and Applications*,
Advances in Intelligent Systems and Computing 291,
DOI: 10.1007/978-3-319-07596-9_8,

processing techniques, such as Chan-Vese segmentation [17], which has excelled in many other image separation tasks, but fails with the type of data involved in this work.

The Convolutional Neural Network (CNN) [14][37] was introduced as a general solution to the image recognition problem of variable inputs. A CNN consists of a multi-layer artificial neural network with a built-in feature extraction process and translational tolerance of the input image space. Therefore, this type of network is capable of accurately identifying images of a target object among cluttered background noise. This is achieved by learning the distinctive features that characterize the class the object belongs to, regardless of the relative position at which it appears in the input image sample.

The article is structured as follow: the next section describes the general background of livestock detection, UAVs and CNN. Section 3 is a system overview in which the system components and the visual recognition techniques used are described in detail. Finally, the last section shows the results and conclusions obtained.

2 Background

There are three areas that converge in this study: (i) livestock species detection and accounting, which is the main objective of the study; (ii) multirotor systems, such as hardware tools used as part of the counting process, in this case the UAV; (iii) image processing techniques used for carrying out the analysis of the images obtained by the UAV, in this case, Convolutional Neural Network. This section describes the current state of the art of each technique. The next section explains how they have been used in this study.

2.1 Livestock Detection

Species accounting is of great interest in activities such as hunting, biology and agriculture. There have been several studies on cynegetic use [23][3] that take into account the advantages [26] and disadvantages [1] with regard to the environment, the economy and tourism.

The difficulty of this task arises from the diversity of soils, species-specific characteristics, and spatial aggregation of animals in the field [15]. In this sense, several authors have made various proposals to solve this problem based on statistical and biological methods [45][25][7].

Thanks to advances in computer science, the task of counting animals can be performed automatically (as opposed to a manual count), as has been addressed successfully in several studies [24][2][38].

2.2 Multirotor System

One of the most popular technological advances in recent years is the multirotor or multicopter, a type of UAV capable, among other characteristics, of aerial filming as described in [39]. The use of this system to obtain an aerial video with sufficient quality is a tool that, in combination with image analysis to detect animals, can save considerable time and money in the process of counting animals.

To carry out the case study transmission of large data from multirotors, it is necessary to have a powerful computer that can perform real time processing. The transmission of large amounts of data requires at least one channel with high bandwidth. Most multirotor systems that exist today use analogue channels to transmit flight orders to the multirotor and send the captured video [47][20][46][6][41].

The proposed system presents an alternative to these systems by taking advantage of technological advances. It tries to take advantage of powerful Wi-Fi antennas currently available in the market, to unify the process of sending information through a single digital channel, including flight commands, telemetry and video in high resolution. Currently there are some systems that utilize UAV communication using sockets [4], but their use is restricted to UAV type aircraft.

This form of communication makes it possible to control the multirotor as well as to obtain information about the status of the sensors it can carry, such as the Global Positioning System (GPS) or altimeter. Of course, it can also send video images taken by its camera. Furthermore, the use of a computer instead of a radio station as a multirotor control element allows a processing capacity that the station does not have, which makes the computer able to control the multirotor intelligently without a human pilot.

2.3 Convolutional Neural Network

The layer architecture for the CNN used will vary according to the application. However, the count of output neurons in the final layer will always be associated with two output classes: one for the object of interest, and the other for background noise. Hence, every time the network is executed over an image patch, it will output two values, each of which will be interpreted as the confidence level with which the network believes the corresponding class correctly describes the analyzed sample.

The CNN is trained with data manually collected from previous UAV test flights, and later artificially augmented. The augmentation consists of increasing the number of training data available by applying a series of transformations to translate, rotate and scale each manually collected example. Such a process can yield up to 40 artificial samples for every original image patch, thus giving the network a lot more data to train with. The training data is separated into two sets for each of the classes, target and background.

The network is finally trained with the prepared data sets through the stochastic gradient descent method for back-propagation [5], which offers an optimal route to minimizing the classification error for this type of highly mutable training data

3 System Overview

The architecture of the multirotor for obtaining images consists of three main parts: hardware, software and communication protocol.

The hardware part refers to the system running the software and is present at both ends of the communication. At one end is the UAV, a multirotor with 6 engines and 6 arms bearing a Single Board Computer (SBC) with a 700MHz processor and 512MB memory capable of running a Linux distribution. The connectivity of the SBC allows connecting a camera via Ethernet and a Wi-Fi antenna Universal Serial Bus (USB); it also gets the information that Speed Electronic Controllers (ESCs) need to control the multirotor motors. At the other end of the communication is an access point whose potency depends on the distance to be achieved, and can cover distances of up to 3 kilometers. This access point will connect the SBC UAV control computer with a high performance laptop with a USB gamepad connected for manual control.

Regarding the software, SBC runs a piece of software that mainly reads, processes and delivers telemetry. It is responsible for carrying out the calculations for the behavior and stability of the multirotor. For its part, the computer runs a piece of software specifically developed (in Java) to control the multirotor both manually and autonomously through waypoints. The user can see all the information from the sensors that the multirotor sends in real time, such as information relating to power consumption, intensity and quality of the Wi-Fi signal, image and video (Fig. **3**). This information will generate log files in XML format that will serve to make a more accurate and detailed analysis of the captured images, as it will provide additional useful information such as height and GPS.

Finally, the communication protocol between the two parts is based on a multi-agent platform called PANGEA (Platform for Automatic coNstruction of orGanizations of intElligent Agents) [48] based on the IRC protocol. Many previous studies [9][13][31][30] support the use of multi-agent systems [12][34][33][10] combined with techniques such as neural networks as an optimal solution to similar problems but in other fields of research [32][22][27][36][28][40][29][21][11][8][35]. With the use of this platform, described in [48], each of the parties involved in the communication will behave as an agent following the proposed PANGEA scheme. In this way, the system is ready for a possible extension where one computer can control different multirotors at the same time and multirotors can communicate with each other to achieve common goals more efficiently [16][44][42][43][19][18].

3.1 Runtime Operation

At runtime, object recognition is carried out by analyzing individual frames received from an on-board camera. A sliding window approach is taken to sequentially analyze small, adjacent, and overlapping image patches positioned in a grid pattern over the frame. Each of these image patches is evaluated by the trained network, and the output values are recorded.

This process is repeated at three scales. The middle scale is chosen to approximate the relative size at which the target objects are expected to be seen – a value which can be calculated trigonometrically based on the current flight altitude of the UAV. The two remaining scales are set at 85% and 115% of the middle scale window size. These additional scales provide supplementary information which later helps to boost the readings obtained at each grid coordinate.

For every window analyzed, the output values of the CNN are processed with the softmax activation function described in Equation 1, where y_i is the resulting value of the network for output neuron *i*. This transformation results in a probability-like value *P(i)*, which for any given class *i* estimates the likelihood that the analyzed window belongs to it.

$$P(i) \equiv \sigma(y,i) = \frac{e^{y_i}}{\sum_{n=0}^{N} e^{y_n}} \tag{1}$$

Equation 2 defines L_{xys} as the likelihood value that the analyzed window at grid coordinate *(x;y)* and scale s belongs to the target class. Combining these values over the entire grid for every x, y and s will lead to a discrete 2D probability distribution over the original input frame that will indicate the positions at which a target object has been detected.

Boosting the values at each coordinate through a simple nearest-neighbor clustering algorithm, where the likelihood value for each point in the grid is increased by strong adjacent readings, can further enhance this distribution but left unchanged by a lack thereof. This process is detailed in Equation 3.

$$L_{xys} \equiv P(target|x,y,s) \tag{2}$$

$$B_{xy} = \sum_{i=x-1}^{x+1} \sum_{j=y-1}^{y+1} \sum_{s=0}^{2} \begin{cases} L_{ijs} & if\ L_{ijs} \geq 1/2 \\ 0 & if\ L_{ijs} < 1/2 \end{cases} \tag{3}$$

The boosted value B_{xy} can then passed through a threshold, determined ad hoc for the particular application, to finally produce a quantized representation of the probability distribution.

An additional benefit of boosting involves the elimination of most false positives, occurrences from which neural networks are never exempt, but which, fortunately, often tend to appear without any neighboring support, thus disappearing after applying this process. Likewise, the use of this mechanism will make the system more tolerant towards false negatives, as these will not have such a significant impact on the final result, due to the network often finding reinforcing values at adjacent positions.

3.2 Visual Recognition

For this application, the network utilized follows an architecture given by 64x64-18C7-MP4-96C5-MP2-4800L-2, wherein there are two convolutional feature extraction stages and one hidden linear layer. The training data is prepared as previously described, with the *target* class consisting of individual animal samples. A small subset of this data set can be seen in Fig. 1.

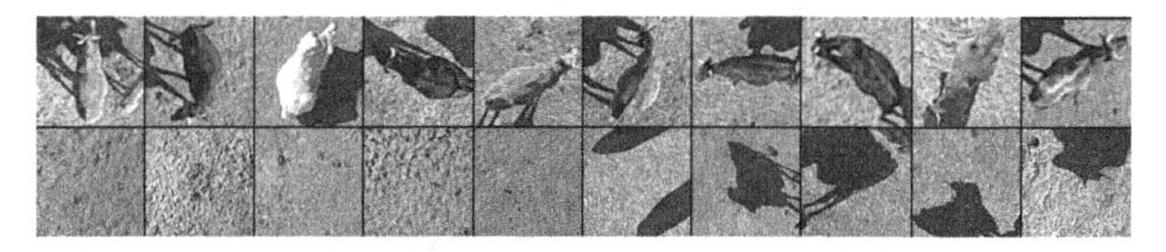

Fig. 1. A subset of the training data used for the CNN, where the dataset is divided into two classes: target (top row) and background (bottom row)

Fig. 2 depicts the runtime implementation of the system, where the multiple stages of detection can be seen. The final quantized likelihood distribution can be used to easily count the cattle currently in view.

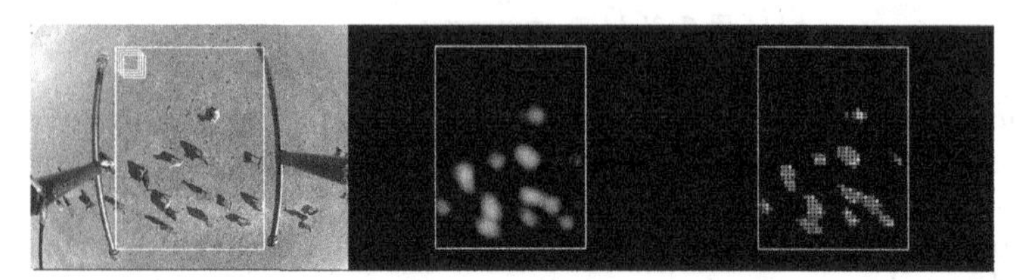

Fig. 2. A sample video frame (left), analyzed with the CNN over a grid pattern producing a probability distribution Lxys plotted over the entire frame (middle), values which can then be boosted and quantized as B_{xy} to better distinguish individual targets (right)

4 Results and Conclusions

This study was developed to test software that controls multirotors and allows visualization of all information captured from the air in real time. It also has a tool that allows for autonomous routes based on waypoints. The multirotor is equipped with two cameras: a flight chamber which provides guidance for the pilot who is controlling the UAV, and a second high resolution camera placed perpendicular to the images obtained from counting the species. Further, the CNN used as a counting technique produced good results, as shown below.

Fig. 3. Ground Control Station software during a flight

The CNN can be trained surprisingly fast, reaching a plateau for the training criterion (MSE on classification) after only a few SGD epochs. A summary of the CNN training results is given in Table 1 Confusion matrix of the visual classification CNN trained with the augmented sample data over 10 SGD epochs..

Table 1. Confusion matrix of the visual classification CNN trained with the augmented sample data over 10 SGD epochs

	Training Data Classification 10,816 Samples			**Testing Data Classification** 2,704 Samples		
Class	Cattle	Background	Accuracy	Cattle	Background	Accuracy
Castle	**5,202**	206	96.2%	**1,241**	111	91.8%
Background	109	**5,299**	98.0%	10	**1,342**	99.3%
Global			**97.1%**			**95.5%**

We can obtain an animal count for every analyzed frame by running a connected component-labeling algorithm over the boosted and thresholded B_{xy} values in that image. The results of this process applied on evenly spaced frames from a 3 minute video sequence are reported in Table 2.

Table 2. Results of the cattle counting application over 70 sample frames analyzed. They are divided according to the counting result, which makes it evident that most errors occur in overcrowded frames, when there is an average of 10 or more visible targets.

Result	Frames analyzed	Targets total	Targets per frame	Targets counted	Δ Count total	Δ Count per frame	Δ Count per target
Over-count	2	22	11.00	24	+2	+1.00	0.091
Correct	57	309	5.42	309	0	0.00	0.000
Undercount	11	112	10.18	99	-13	-1.18	0.116
Total	70	443	6.32	432	-11	-0.16	**0.025**

Finally, it should be noted that the use of PANGEA as the platform for communication between the computer and the multirotor opens new research lines since PANGEA provides the capacity and mechanisms for several multirotors to communicate simultaneously. This means that it will be possible to explore areas in an autonomous and collaborative way [18], allowing the task of flying over the area to be distributed among the various required multirotors, thus reducing the flight time or increasing the area covered.

Acknowledgements. This work has been carried out by the project *Sociedades Humano-Agente: Inmersión, Adaptación y Simulación*. TIN2012-36586-C03-03. Ministerio de Economía y Competitividad (Spain). Project co-financed with FEDER funds.

References

[1] San Miguel, A., Ochoay, J., Pérez Carral, C.: Wildlife Management in Mediterranean Forest Ecosystems. Problems and current situation of Shrublands. Montes (35), 33–36 (1994)

[2] B. T. a. C. J.: Real-time Face Detection and Tracking of Animals. In: de 8th Seminar of Neural Network Applications in Electrical Engineering, Belgrade (2006)

[3] Buenestado, B.V.: Aprovechamiento y gestión de la caza en España. Una reflexión a propósito de los cercados cinegéticos. Actas del VI Coloquio de Geografía Rural, 257–272 (1991)
[4] Bodin, W.K., Redman, J.J., Thorson, D.C.: U.S. Patent No. 7,286,913. U.S. Patent and Trademark Office, Washington, DC (2007)
[5] Bottou, L.: Stochastic learning. In: Bousquet, O., von Luxburg, U., Rätsch, G. (eds.) Machine Learning 2003. LNCS (LNAI), vol. 3176, pp. 146–168. Springer, Heidelberg (2004)
[6] Buss, H., Busker, I.: Mikrokopter (2008)
[7] C. S. a. R. T. J. McKinlay: Integrating Count Effort by Seasonally Correcting Animal Population. CCAMLR Science 17, 213–227 (2010)
[8] Pinzón, C.I., Bajo, J., De Paz, J.F., Corchado, J.M.: S-MAS: An adaptive hierarchical distributed multi-agent architecture for blocking malicious SOAP messages within Web Services environments. Expert Systems with Applications 38(5), 5486–5499
[9] Tapia, D.I., Abraham, A., Corchado, J.M., Alonso, R.S.: Agents and ambient intelligence: case studies. Journal of Ambient Intelligence and Humanized Computing 1(2), 85–93 (2010)
[10] Tapia, D.I., De Paz, J.F., Rodríguez, S., Bajo, J., Corchado, J.M.: Multi-agent system for security control on industrial environments. International Transactions on System Science and Applications Journal 4(3), 222–226 (2008)
[11] Tapia, D.I., Alonso, R.S., De Paz, J.F., Corchado, J.M.: Introducing a distributed architecture for heterogeneous wireless sensor networks. In: Omatu, S., Rocha, M.P., Bravo, J., Fernández, F., Corchado, E., Bustillo, A., Corchado, J.M. (eds.) IWANN 2009, Part II. LNCS, vol. 5518, pp. 116–123. Springer, Heidelberg (2009)
[12] Tapia, D.I., Rodríguez, S., Bajo, J., Corchado, J.M.: FUSION@, a SOA-based multi-agent architecture. In: International Symposium on Distributed Computing and Artificial Intelligence 2008 (DCAI 2008), pp. 99–107 (2008)
[13] Fdez-Riverola, F., Corchado, J.M.: CBR based system for forecasting red tides. Knowledge-Based Systems 16(5), 321–328 (2003)
[14] Fukushima, K.: Neocognitron: A self-organizing neural network model for a mechanism of pattern recognition unaffected by shift in position. Biological Cybernetics 36(4), 193–202 (1980), doi:10.1007/BF00344251
[15] Sileshi, G.: The excess-zero problem in soil animal count data and choice of appropriate models for statistical inference. Pedobiologia 52(1), 1–17 (2008)
[16] Garijo, F., Gómes-Sanz, J.J., Pavón, J., Massonet, P.: Multi-agent system organization: An engineering perspective. In: Pre-Proceeding of the 10th European Workshop on Modeling Autonomous Agents in a Multi-Agent World (MAAMAW 2001) (2001)
[17] Getreuer, P.: Chan-Vese Segmentation. Image Processing on Line 2012 (2012), doi:10.5201/ipol.2012.g-cv
[18] Gómez, J., Patricio, M.A., García, J., Molina, J.M.: Communication in distributed tracking systems: an ontology-based approach to improve cooperation. Expert Systems 28(4), 288–305 (2011)
[19] Griol, D., García-Herrero, J., Molina, J.M.: Combining heterogeneous inputs for the development of adaptive and multimodal interaction systems. Advances in Distributed Computing and Artificial Intelligence Journal 6, 37–53 (2013) ISSN 2255-2863
[20] Haehnel, H.: Remote controlled flying robot platform. In: Third International Conference on Digital Information Management, ICDIM 2008, pp. 920–921. IEEE (November 2008)
[21] Bajo, J., De Paz, J.F., Rodríguez, S., González, A.: Multi-agent system to monitor oceanic environments. Integrated Computer-Aided Engineering 17(2), 131–144 (2010)

[22] Bajo, J., Corchado, J.M.: Evaluation and monitoring of the air-sea interaction using a CBR-Agents approach. In: Muñoz-Ávila, H., Ricci, F. (eds.) ICCBR 2005. LNCS (LNAI), vol. 3620, pp. 50–62. Springer, Heidelberg (2005)
[23] J. C. G. R. R. A. B. a. J. M. V. M. Ángel Farfán: Game harvest characterisation of the mammals in Andalusia. Galemys 16(1), 41–59 (2004)
[24] J. D. R. a. K. R. G. Felix A. Wichmann: Animal detection in natural scenes: Critical features revisited. Journal of Vision 10(4) (April 2010)
[25] Anderson, J.I.J.M.: Tropical Soil Biology and Fertility. A Handbook of Methods. de CAB International, Wallingford (1993)
[26] Gallego, J.I.R.: Caza y turismo cinegético como instrumentos. Anales de Geografía de la Universidad Complutense 30(2) (Octubre 2010)
[27] Fraile, J.A., Bajo, J., Corchado, J.M., Abraham, A.: Applying wearable solutions in dependent environments. IEEE Transactions on Information Technology in Biomedicine 14(6), 1459–1467 (2011)
[28] De Paz, J.F., Rodríguez, S., Bajo, J., Corchado, J.M.: Case-based reasoning as a decision support system for cancer diagnosis: A case study. International Journal of Hybrid Intelligent Systems 6(2), 97–110 (2009)
[29] De Paz, J.F., Rodríguez, S., Bajo, J., Corchado, J.M.: Mathematical model for dynamic case-based planning. International Journal of Computer Mathematics 86(10-11), 1719–1730 (2009)
[30] Corchado Rodríguez, J.M.: Redes Neuronales Artificiales: un enfoque práctico. Servicio de Publicacións da Universidade de Vigo, Vigo (2000)
[31] Corchado, J.M., Lees, B.: Adaptation of cases for case based forecasting with neural network support. In: Soft Computing in Case Based Reasoning, pp. 293–319 (2001)
[32] Corchado, J.M., Fyfe, C.: Unsupervised neural method for temperature forecasting. Artificial Intelligence in Engineering 13(4), 351–357 (1999)
[33] Corchado, J.M., Aiken, J., Rees, N.: Artificial intelligence models for oceanographic forecasting. Plymouth Marine Laboratory (2001)
[34] Corchado, J.M., Aiken, J.: Hybrid artificial intelligence methods in oceanographic forecast models. IEEE Transactions on Systems, Man, and Cybernetics, Part C: Applications and Reviews 32(4), 307–313 (2002)
[35] Corchado, J.M., Bajo, J., De Paz, J.F., Rodríguez, S.: An execution time neural-CBR guidance assistant. Neurocomputing 72(13), 2743–2753 (2009)
[36] Corchado, J.M., De Paz, J.F., Rodríguez, S., Bajo, J.: Model of experts for decision support in the diagnosis of leukemia patients. Artificial Intelligence in Medicine 46(3), 179–200 (2009)
[37] LeCun, Y., Bottou, L., Bengio, Y., Haffner, P.: Gradient-based learning applied to document recognition. Proceedings of the IEEE 86(11), 2278–2324 (1998), doi:10.1109/5.726791
[38] Parihk, M., Pately, M., Bhat, D.: Animal Detection Using Template Matching Algorithm. International Journal of Research in Modern Engineering and Emerging Technology 1(3) (2013)
[39] Mahony, R., Kumar, V., Corke, P.: Multirotor aerial vehicles: Modeling, estimation, and control of quadrotor (2012)
[40] Borrajo, M.L., Baruque, B., Corchado, E., Bajo, J., Corchado, J.M.: Hybrid neural intelligent system to predict business failure in small-to-medium-size enterprises. International Journal of Neural Systems 21(4), 277–296 (2011)
[41] Ramli, H., Kuntjoro, W., Makhtar, A.K.: Advanced Autonomous Multirotor Response System. Applied Mechanics and Materials 393, 299–304 (2013)

[42] Rodriguez, S., Julián, V., Bajo, J., Carrascosa, C., Botti, V., Corchado, J.M.: Agent-based virtual organization architecture. Engineering Applications of Artificial Intelligence 24(5), 895–910

[43] Rodríguez, S., Pérez-Lancho, B., De Paz, J.F., Bajo, J., Corchado, J.M.: Ovamah: Multiagent-based adaptive virtual organizations. In: 12th International Conference on Information Fusion, FUSION 2009, pp. 990–997 (2009)

[44] Rodríguez, S., de Paz, Y., Bajo, J., Corchado, J.M.: Social-based planning model for multiagent systems. Expert Systems with Applications 38(10), 13005–13023 (2011)

[45] Macrofauna, S., Lavelle, P., Senapati, B., Barros, E.: Trees, Crops and Soil Fertility: Concepts and Research Methods, pp. 303–323 (2003)

[46] Svanfeldt, M.: Design of the hardware platform for the flight control system in an unmanned aerial vehicle (Doctoral dissertation. Linköping) (2010)

[47] Tretyakov, V., Surmann, H.: Hardware architecture of a four-rotor UAV for USAR/WSAR scenarios. In: Workshop Proceedings of SIMPAR 2008-International Conference on Simulation, Modeling and Programming for Autonomous Robots (2008)

[48] Zato, C., Villarrubia, G., Sánchez, A., Bajo, J., Corchado, J.M.: PANGEA: A New Platform for Developing Virtual Organizations of Agents. International Journal of Artificial Intelligence TM 11(A13), 93–102 (2013)

Easy Development and Use of Dialogue Services

José Javier Durán[1], Alberto Fernández[1], Sara Rodríguez[2], Vicente Julián[3], and Holger Billhardt[1]

[1] CETINIA, University Rey Juan Carlos, Móstoles, Spain
{josejavier.duran,alberto.fernandez,holger.billhardt}@urjc.es
[2] Universidad de Salamanca, Salamanca, Spain
srg@usal.es
[3] Universidad Politécnica de Valencia, Spain
vinglada@dsic.upv.es

Abstract. We present a framework for Dialogue-Based Web Services (DBWS), i.e. services that require several message exchanges during their execution. Service development is simplified with the use of script languages and abstracting the communication layer. Service advertisements are carried out with a semantic Web Service directory with search and reputation capabilities. Execution can be performed from a mobile user interface that includes capabilities for user assistance. Our framework aims at filling the gap between services and non-IT users/experts. An example illustrates our proposal.

Keywords: Web services, Service directory, Middleware, User assistance.

1 Introduction

When humans request support from experts in some field, they do not usually exchange a single message with the problem description and an answer/solution from the expert. However, they typically engage in several interactions where the expert asks for context information, desires, etc., where questions may depend on previous answers and expert knowledge. The same approach should apply when one (or several) of the previous roles (usually the expert) are played by software agents.

Building such software systems is not an easy task. Even though many experts are able to program software pieces (knowledge bases) like rule-based, logic, scripts, etc. they usually lack skills to create software accessible by humans or agents (Web applications, Web services, software agents, ...).

An additional problem is how a user can access those services. Firstly, the user needs to find a service that might be of interest. Then, the service has to be used, i.e. invoked passing the necessary parameters, possibly requiring several interactions as mentioned above.

In this paper we propose a framework focused on filling the gap between Web Services (WS) and humans, at different levels. First, the framework supports the development of WS using different scripting languages, and isolates the communication layer associated to WS from the dialogue process. Next, services are indexed in a

C. Ramos et al. (eds.), *Ambient Intelligence - Software and Applications*,
Advances in Intelligent Systems and Computing 291,
DOI: 10.1007/978-3-319-07596-9_9, © Springer International Publishing Switzerland 2014

directory capable of searching services using different techniques (including free text). Also, that directory enriches search results with reputation information, in order to assist users to choose the most reliable/best service, based on other user's experiences. Finally, a generic interface is provided for service invocation, which covers mobile devices and offers an assistant that helps users with context information.

The rest of the paper is organized as follows. In section 2 we analyze other related works. Section 3 describes the architecture of our framework. Section 4 explains the development support middleware. An example of using our framework is described in section 5. We finish with conclusions and future works.

2 Related Work

Description languages and transport protocols are important parts of Web services development. There are two main technologies: REST services with JSON payload (mainly described using WADL), and SOAP (as WSDL services). The former is lightweight, easier for developers to understand, and more adaptable. The latter is more widely adopted in industry due to existing standards (WS-*) and tools [1-2]. Deployment environment is another important aspect in the development of Web services. Nowadays industry is moving towards PaaS (Platform as a Service) environments [3] in which different applications are deployed together sharing resources and its highly useful when different applications share a common structure and/or they are used in the same way (e.g. Heroku platform is running more than 3 million applications[1]).

There are different solutions focused on the creation of dynamic interfaces for Web services. Usually, the user interface is created depending on the type of service to use, or the parameters required for its execution. Some of these solutions translate a WSDL description into a Web interface that represents the different kinds of restrictions and input types using HTML widgets [4]. Others are focused on testing services by creating requests based on service definitions, but offering an interface more appropriate to software developers [5]. There are other options that integrate both a directory of services with a test user interface for such services, even including options for user feedback. In particular, there are several existing public service directories. In Table 1 we enumerate the different characteristics that we think should be present in a Web Service directory, and how they are implemented in different solutions. The first characteristic is whether the directory provides *search* capabilities. *Registry* defines whether users can register their own services or the directory is closed. A useful information for selecting services is *reputation*. There are different mechanisms for reputation, such as: rating, users' feedback as comments, or wiki-like in which users can update the description of a service in order to correct any wrong information. By *execution* we mean if it is possible to invoke the service directly from the directory web interface, without needing to develop an ad-hoc application, or if there is specific documentation of that process (e.g. example script, or unitary tests of the service). Finally, *format* represents the kind of services that can be registered (SOAP/WSDL, REST, …).

[1] https://blog.heroku.com/archives/2013/4/24/europe-region

Table 1. Comparison of different web service directories

	Search	Registry	Reputation	Execution	Format
Membrane SOA registry	No (list)	Yes	Rating	Yes Low-level	SOAP
WS-index.org	Text	No	Rating	No	*Unknown*
API-Hub	Text + Filters	Yes	No	No	Any
Programmable web	Text + Filters	Yes	Rating	No	Any
X Methods	No (list)	Yes	No	No	SOAP
BioCatalogue	Text + Filters + In/Out	Yes	No (wiki)	Examples	SOAP, REST
Embrace	Text	No	Comments	Unitary tests	SOAP, REST, DAS, BioMOBY

Membrane SOA Registry[2] includes a five-star rating system and a (low-level) SOAP invocation user interface, but lacks of a search capability. *WS-index.org* is a directory of web-pages related to web services, but a standard format is not applied to the entries, and most of the entries are out-dated. *API-Hub*[3] and *Programmable Web*[4] focus on API documentation and both offer text and filter-based search. *X Methods*[5] offers a WSDL-only directory, but it lacks of search capabilities and reputation mechanisms. *BioCatalogue*[6] offers a complex search mechanism able to filter by text, tags, and kind of input and/or output, but instead of offering an execution mechanism, it serves as a repository of execution examples. *Embrace*[7] is a specialized directory for medical services (support for domain description formats like DAS and Bio-MOBY), which offers access to unitary tests that are run in background in order to measure the reliability of the services. Despite the existence of all those tools, there is a lack of a solution that integrates all the important Web service mediation characteristics together. *Programmable Web* is the most complete regarding those characteristics, but it does not allow execution, which is only supported by *Membrane*. Moreover, they do not provide support for service development.

---

[2] `http://www.service-repository.com/`
[3] `http://www.apihub.com`
[4] `http://www.programmableweb.com/`
[5] `http://www.xmethods.com/`
[6] `https://www.biocatalogue.org/`
[7] `http://www.embraceregistry.net/`

3 Architecture

Fig. 1 shows our framework architecture. There are three main components: a service directory, a middleware and a Web interface.

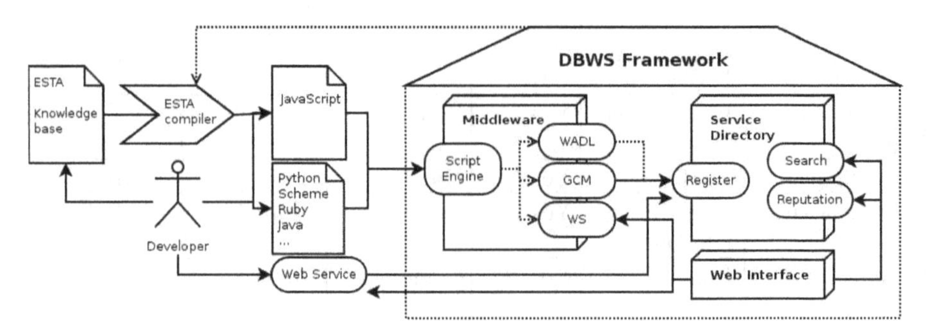

Fig. 1. Framework components

The *Service Directory* acts as a mediator (yellow pages) among services and users. Agents advertise the services they provide by registering with the directory. A service registration includes (i) a *description* of its functionality, (ii) a *grounding* specifying the endpoint where the service can be invoked, and (iii) the agent/organisation that created or owns the service (for reputation management). The service directory coordination is carried out by means of a heterogeneous service directory called Nuwa [6], and reputation management is based on a simplification of the reputation mechanism proposed by Hermoso et al. [7] for task oriented multi-agent systems. In this paper we do not focus on the description of our service directory, which can be found in the references above.

The *Development support middleware* is a set of tools that facilitate the development of dialogue-based services. A *Script Engine* takes script code and generates a Web service implementation (*WS*) and its *GCM* and *WADL* descriptions, as is detailed in next sections. Additionally, the framework includes a compiler to translate *ESTA*[8] knowledge bases into JavaScript code.

The *Web Interface* is a generic Web application that provides a human interface to search and invoke services registered with the directory, as well as providing feedback about service use.

4 Service Development Support Middleware

In order to ease the implementation and integration of Web Services using our framework, we have developed a middleware that deals with process workflow and message exchange. The advantage of this middleware is that it is possible to create a DBWS without implementing any Web functionality, since the communication part is isolated from the application itself. Also, this middleware offers a sandbox

[8] Expert System Shell for Text Animation

environment in which multiple applications can be run together isolated among them, and where errors are properly managed by the middleware.

The main characteristics of the proposed middleware are: (i) isolate the communication layer from the application, (ii) transform Web requests into software objects used by the application, (iii) do not impose a programming language, or paradigm, and (iv) avoid the use of special structures, or patterns, for dialogue management.

4.1 Interaction Protocol

In this section we describe the most important aspects of our framework: a workflow process for dialogue-based services, and a format for message exchange.

Workflow. In order to use dialogue-based services, a record of the interaction has to be kept. Services could be invoked in two states: initialisation and resume. During initialisation a service communicates to the client which parameters must be provided. During resume, the service takes the parameters received and returns a message that may include additional information (parameters) required to continue the execution or the result. The message content is explained next.

Message Format. We divide the dialogue message in three parts:

- *State information*: includes a set of variables representing the service state. This information is used when interacting with stateless services and must be sent to the service again in order to keep a track of the dialogue.
- *Response*: a set of messages that are sent to the client for its use. Each message can be, for example, a text, an HTML document, a picture, or an RDF document. Those messages are considered the output of the service.
- *Question*: When a service requires more information, or asks the user to wait for a time condition to be reached, a question is sent to the client. That question has a textual condition (the question), a motivation (why it is needed, and/or some semantic information about the question), a parameter name (id) (used to send back a client response), and a rule of accepted values (combination of type and values).

4.2 Script Engine Middleware

The script engine relies on the implementation of the JSR-223[9] API present in the Java runtime. This API is capable of loading applications created in different script languages (such as Java, JavaScript, Python, Scheme, Ruby, etc.), offering an abstraction of the communication between Java classes and script applications. The advantage of this approach is that it is possible to access applications independently of the programming language as long as a parser for that language is available.

[9] `http://www.jcp.org/en/jsr/detail?id=223`

4.3 Web Interface for Web Service Invocation

Since our framework defines a common interface for multiple services (the message protocol) it is possible to reuse a user interface to access different services. In our case, we have developed a user interface that covers the main aspects of our proposal: search, invocation, and feedback.

Search. The user interface accesses to the service directory, and offers two kinds of search methods: by keywords or free text. The service directory returns the matching services with their degree of match and reputation. The results are shown to the user ordered by these two parameters. The user can switch between both.

Invocation. The proposed protocol includes information needed for a dialogue stage, i.e. parameter required (question field) and response messages. The user interface shows the response messages followed by the parameter question and by a log of previous responses in the dialogue. The parameter question contains two elements: the parameter question (enriched with motivation information) and the input field. The latter is created with the most appropriate HTML input.

Feedback. During the invocation process, the current reputation score is shown, and the user can submit a feedback about the service. The feedback can include a score, a text about the user's experience and the dialogue log (e.g. for debugging).

In addition to those main functionalities, our Web interface includes a *question assistant* module that provides information related to the current question, e.g. main concept or language translation. The current implementation uses WordReference (synonyms of main terms), and a natural language question answering system (*START*[10]) to clarify the meaning of a concept or even suggest an answer for a question (e.g. if the question is asking about the value of a biochemical parameter, it will offer the textual description of that parameter from Wikipedia.org). Access to Google Translate has been implemented but it is disabled because of its commercial license.

5 Case Study

In this section we use an example to illustrate the process of adapting a specific dialogue-based application to our architecture. We chose a simple application that assists users in deciding what cocktail to make, by asking the user questions about desired ingredients or restrictions (e.g. % alcohol). Fig. 2 shows the interactions involved during a cocktail drinks' assistance. Solid arrows represent user to service messages, while dashed arrows represent service to user ones.

First, the server asks the user which is the limit of alcohol that he wants in the drink (*an enumeration*). The user answers '*< 50%*'. For clarity, we simplified the question field, omitting their description. Then, the service asks whether the user wants it with some juice (*true/false*) and the user answered *true*. Next, Vodka is offered as a possible ingredient (*true/false*), and the user agrees (*true*). Finally, the

[10] `http://start.csail.mit.edu/`

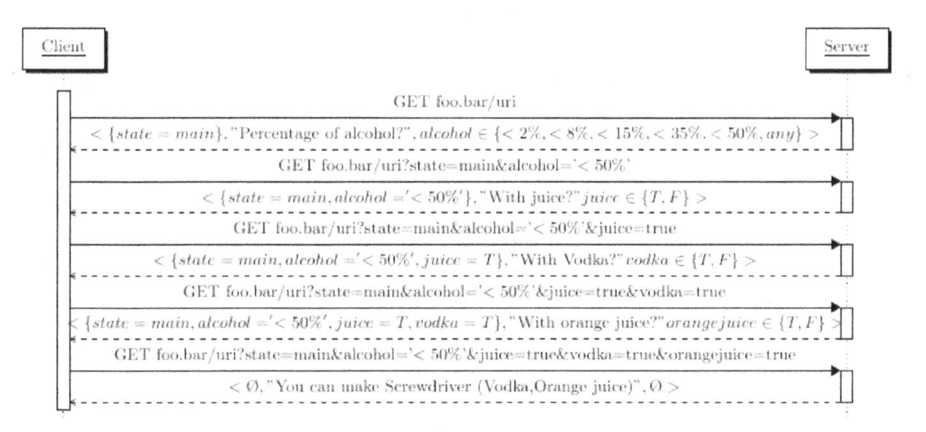

Fig. 2. HTTP message exchange between a web client and the service. Message format: <state information, response, question>

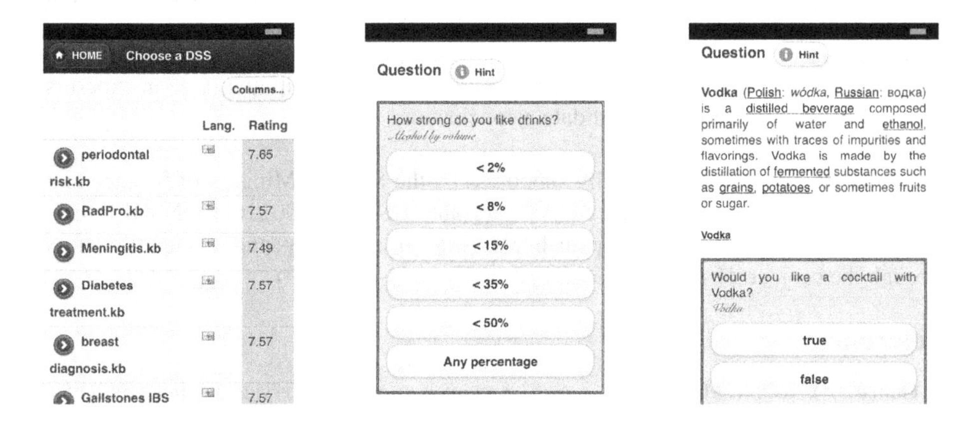

Fig. 3. Mobile Web User Interface for the application

service asks the user if he also wants something with orange juice (*true*/*false*), which he agrees (*true*). As a result, a cocktail is found in the knowledge base, and the service closes the dialogue with the recipe as a message with no further values to be provided.

Fig. 3 shows several snippets of the user interface, in particular obtained from a mobile phone access to the service. The first one shows the selection screen, in which a DBWS can be chosen. Next screens show questions 1 and 3 from the previous sequence diagram, including information from the question assistant (Vodka definition).

6 Conclusion

In this paper we have described a framework for developing and interacting with Dialog-Based Web Services. The main contributions of this paper are (i) a framework that supports service development by providing an integration component for different scripting languages, which definitely facilitates Web service implementation; and (ii) a generic Web interface that supports the user to invoke such services. The framework includes a multi-language service directory with registration, search and reputation mechanisms, which we adopted from previous work.

Although our framework includes several components (Directory, Script Engine, Web Interface), developers can use their desired functionalities. Then, they might want to use only the directory functionality by registering their services. Or they might want to provide a script (or ESTA) implementation to the Script Engine so as to generate the Web service. Independently of the previous options, the Web Interface tool can be used to search and/or (v) invoke services if wanted.

The proposed framework has been implemented and we are currently working on its use for the development of a system to assist clinicians in their diagnosis. The system integrates 16 different knowledge-based medical decision support systems. Those systems are programmed in *ESTA* expert system, and have been integrated in our framework in straightforward way. We use the user interface presented in this paper to test that system. We will use that application to evaluate our framework in a real case, including the reputation mechanism with feedback provided by domain experts (clinicians).

In the future, we also plan to extend our approach to deal with asynchronous services, i.e. services that must pause their execution and resume it later (e.g. an expert is required to emit a response, or validate a conclusion).

Acknowledgement. Work partially supported by the Spanish Ministry of Science and Innovation through the project "AT" (grant CSD2007-0022; CONSOLIDER-INGENIO 2010) and by the Spanish Ministry of Economy and Competitiveness through the project iHAS (grant TIN2012-36586-C03-01/02/03).

References

1. Guinard, D., Ion, I., Mayer, S.: In search of an internet of things service architecture: REST or WS-*? A developers' perspective. In: Puiatti, A., Gu, T. (eds.) MobiQuitous 2011. LNICST, vol. 104, pp. 326–337. Springer, Heidelberg (2012)
2. Pautasso, C., Zimmermann, O., Leymann, F.: Rest- ful web services vs. "big"' web services: Making the right architectural decision. In: Proceedings of the 17th International Conference on World Wide Web, WWW 2008, pp. 805–814. ACM (2008)
3. Lawton, G.: Developing software online with platform-as-a- service technology. Computer 41(6), 13–15 (2008)
4. Kopel, M., Sobecki, J., Wasilewski, A.: Automatic web-based user interface delivery for soa-based systems. In: Bădică, C., Nguyen, N.T., Brezovan, M. (eds.) ICCCI 2013. LNCS, vol. 8083, pp. 110–119. Springer, Heidelberg (2013)
5. Bartolini, C., Bertolino, A., Marchetti, E., Polini, A.: Ws-taxi: A wsdl-based testing tool for web services. In: International Conference on Software Testing Verification and Validation, ICST 2009, pp. 326–335 (2009)
6. Fernandez, A., Cong, Z., Balta, A.: Bridging the gap between service description models in service matchmaking. Multiagent and Grid Systems 8(1), 83–103 (2012)
7. Hermoso, R., Billhardt, H., Centeno, R., Ossowski, S.: Effective use of organisational abstractions for confidence models. In: O'Hare, G.M.P., Ricci, A., O'Grady, M.J., Dikenelli, O. (eds.) ESAW 2006. LNCS (LNAI), vol. 4457, pp. 368–383. Springer, Heidelberg (2007)

Design and Implementation of the Intelligent Plant Factory System Based on Ubiquitous Computing

Jeonghwan Hwang, Hoseok Jeong, and Hyun Yoe*

Department of Information and Communication Engineering,
Sunchon National University, Maegok-dong, Suncheon-si, Jeollanam-do, Korea
{jhwang,hsjeong,yhyun}@sunchon.ac.kr

Abstract. Abnormal climate events have caused agricultural production decrease and its supply insecurity. These problems, coupled with population growth, have led to food shortage. As a solution to such a situation, plant factories capable of producing farm products regardless of climate or location have garnered broader attention. In this research, we propose a ubiquitous computing-based intelligent plant factory system designed by improving the previous plant factory system with just a simple function of On/Off. The proposed system collects information on necessary environmental factors for plants growth, nutrient solution, etc. through sensors installed in a plant factory and infers contextual information through the ontology based on the collected data to provide corresponding appropriate services to users. In addition, this system supports various application platforms such as web PDA, smart phones, etc. so that users can access its service on the plant factory anywhere and anytime.

Keywords: Agriculture, Plant Factory, Ubiquitous Computing, Intelligent System.

1 Introduction

Global warming and environmental pollution of recent days have brought abnormal climate events more frequently such as heat wave, drought, and localized heavy rain. Such irregular natural environmental changes have affected crop growth to lower agricultural production and disturb its stable supply [1].

These problems can cause food shortage amid population growth and possibly lead to food crisis. As a way to resolve this issue, plant factories for crop protected cultivation have received great attention [2].

A plant factory refers to a certain controlled system for crop cultivation where environmental conditions are artificially controlled such as light, temperature, humidity, CO_2 and culture medium, in order to produce crops regardless of season or location[3]. A plant factory, with its function to precision control the artificial conditions for crop growth, can contribute to improved productivity by producing crops

* Corresponding author.

C. Ramos et al. (eds.), *Ambient Intelligence - Software and Applications*,
Advances in Intelligent Systems and Computing 291,
DOI: 10.1007/978-3-319-07596-9_10,

throughout the years and high value-added farming by making it possible to farm functional crops hard to outdoors [4].

Plant factories started in northern Europe in the 1960s and relevant researches have been continued centering on advanced countries such as US, Japan, Canada, and Netherlands since the 1980s [5, 6]. In South Korea, starting from the nutrient solution culture experiment by the agricultural engineering research institute of Rural Development Administration in 1996, plant factory system was established in 2005 to operate an experiment center, almost rivaling with other advanced states [7, 8]. But more studies on precision control of environmental factors are still necessary for plant factories [9].

In this research, we propose a ubiquitous computing-based intelligent plant factory system by improving the previous system that monitors its crop growth environment and simply On/Off controls the actuator according to the monitored data.

The proposed system collects environmental data through temperature, humidity, illumination intensity, CO_2, EC and pH sensors installed in a plant factory and infers contextual information through the ontology based on the collected environmental data to control the actuator automatically and appropriately for each situation and maintain the optimal crop growth environment.

Recently, Web Service has been studied in various fields such as e-commerce, web security, online banking, shopping, e-government, etc [10, 11, 12]. We developed web applications for PC using this web service and mobile application for smartphone based on Android OS. In the research, developed applications support that users can monitor and control their plant factories anytime and anywhere.

The structure of this research is as follows; Chapter 2 explains the structure of the proposed intelligent plant factory. Chapter 3 presents the results of our implementation of the proposed system. Lastly, Chapter 4 summarizes and concludes this research.

2 Design of the Proposed Intelligent Plant Factory System

For the purpose to improve the previous plant factory system that only simply controls the actuator, the proposed intelligent plant factory system utilizes the ontology to realize more intelligent environmental control. The proposed system is also operable in various application platforms to realize plant factory monitoring and control in any place at any time.

Sensors include weather sensors collecting information on intra-factory weather conditions and root zone sensors collecting information on nutrient solution that supplies nutrients to crop roots. The weather sensors measure temperature, humidity, illumination intensity, CO_2, and other elements affecting crop growth. The root zone sensors measure EC and pH of supplied liquid, the amount, EC, and pH of waste liquor, rate of absorption and temperature within culture medium, and temperature of supplied water and waste liquor, etc.

CCTVs are installed both inside and outside the plant factory to collect plant-factory video information and crop video information.

The actuator is consisted of a ventilation and cooling/heating system, CO_2, supply system, nutrient supply system and light source supply system which can control the plant environment and nutrient solution. Each system is also controlled through PLC(Power Line Communication).

The management server is consisted of sensor manager managing environmental information collected by sensors, image manager managing video information from CCTVs, actuator manager administering plant factory actuator, an inference engine that infers situation and surrounding environment patterns regarding a certain event by referring to other information from many objects in the collected data and provides suitable service accordingly, and a database saving plant factory environmental information, video information, ontology for contextual awareness, etc.

Sensor manager refines, filters, and converts data from the sensors while saving the converted data in a database by using update inquiry. Image manager provides streaming data to the web by transferring images from the CCTVs, and saves it to the database after classifying it by plant factory ID and camera number. Actuator manager operates or the manages actuator through PLC by receiving control signals from the inference engine or a user application, and conditions of such a actuator, operation time and the number of controls are saved in the database. The inference engine uses the ontology based on the collected data to infer the situation of crop growth environment and growth stages; searches the most appropriate service based on the inferred contextual information and service rules that define the scope of possible services; and triggers an event suitable for the inferred service to provide a proper services for each different situation.

User application provides various services to users through web, PDA and smartphones such as plant factory environment monitoring, video monitoring, and plant factory crop growth environment control.

Figure 1 shows the system architecture for proposed intelligent plant factory system.

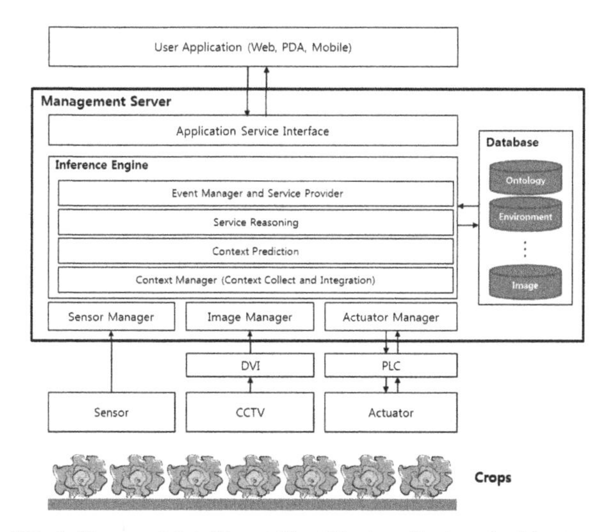

Fig. 1. Proposed Intelligent Plant Factory System Architecture

3 Implementation

3.1 Implementation Ontology for Intelligent Control Service

To provide this intelligent control service for a plant factory based on contextual awareness, an ontology is constructed as shown in Figure; the ontology design uses Protégé [13]. The ontology in this research has 6 upper classes. Network represents information of sensor networks, and Sensor, Node, Location, and Context represent sensor, node, location, and context information, respectively, and Service represents services that users could be provided in the plant factory system. Such information is written as OWL [14] documents and used with JENA [15].

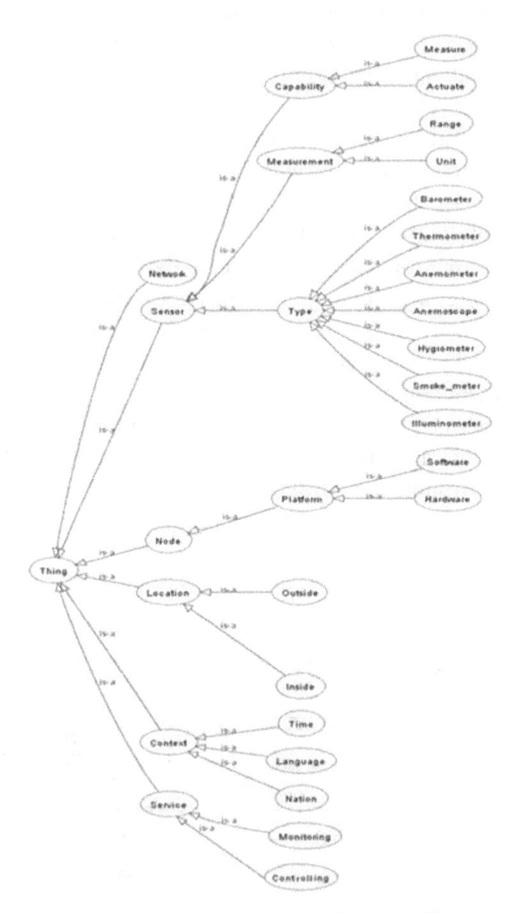

Fig. 2. Intelligent Control Service OWL-Model(Class) for the Proposed System

3.2 Implementation of the Proposed Intelligent Plant Factory System

To verify the performance of the proposed system in this research, we established a plant factory as shown in the figure3. The plant factory was structured in 3-tier bed for the analysis and comparison of the operational performance of established devices.

Fig. 3. Construction of the Proposed Intelligent Plant Factory

In order to gather information about the plant factory environment, sensors were installed inside to measure temperature, humidity, illumination intensity, CO_2 and other weather conditions. In addition, considering that plant factory mainly uses nutrient solution culture, management of the rooting zone, which greatly affects absorption in nutrient solution culture, is very important, and factors such as the amount of nutrients, EC and pH of supplied liquids, the amount, EC, and pH of waste liquors, rate of absorption and temperature within the culture medium, and temperature of water supply and waste liquor, were measured by installing sensors to collect information about the rooting zone environment.

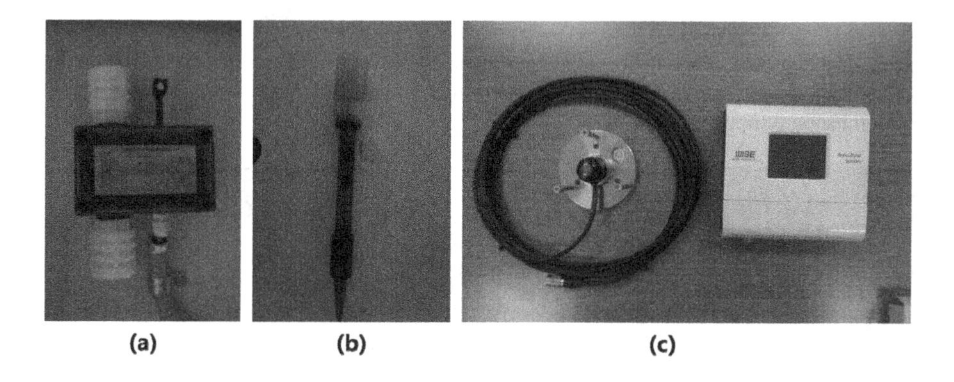

Fig. 4. Installed Sensors in Proposed Intelligent Plant Factory System; (a) Temperature, Humidity and CO2 Sensor, (b) EC and pH Sensor, and (c) illumination intensity Sensor

In order to create an optimized crop growth environment based on the crop growth information, weather information and root environmental information collected from these sensors, we installed the actuator of a ventilation and cooling/heating system, CO_2 supply system, nutrient solution supply system and LED light source supply system inside the factory along with the PCL controller for their control as shown in the figure.

Fig. 5. Installed Actuators in Proposed Intelligent Plant Factory System; (a) ventilation and cooling/heating system, (b) CO_2 supply system, (c) nutrient solution supply system and (d) LED light source supply system.

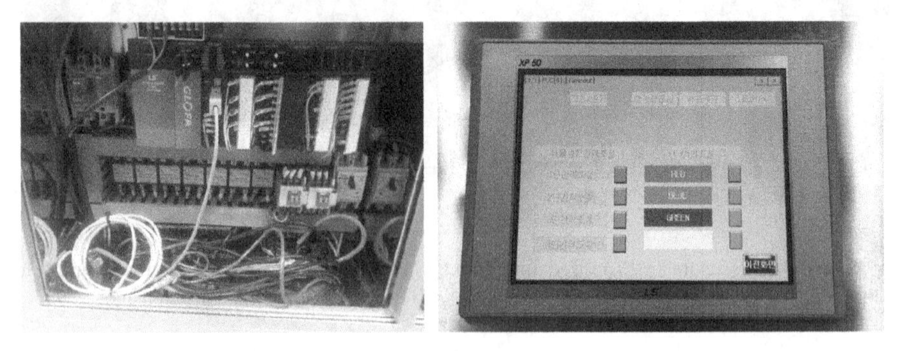

Fig. 6. Installed PLC and Controller in Proposed Intelligent Plant Factory System

In addition, for the real-time monitoring and control of the established plant factory, we was developed a web application for PC and a mobile application.

The web application was developed based on JAVA and C# in Window XP Service Pack3 OS. WAS (Web Application Server) uses Tomcat-6.0.20 and the database Mysql 5.0 which is the safest version among the versions that are currently released. Figure 7 shows the plant factory management GUI of the implemented web application.

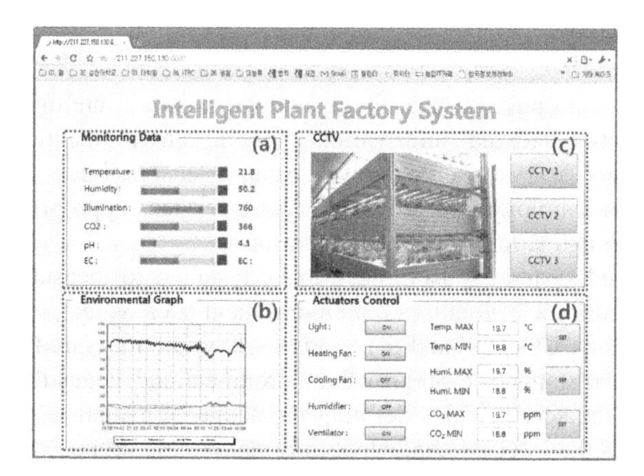

Fig. 7. Web Application GUI

In the plant factory management GUI, the sensing values measured by the sensors installed plant factory appeared in (a), and (b) shows a graph of average values of the data from the sensors; (c) presents the streaming video data received from the cameras installed in plant factory; and (d) shows control of actuators in plant factory and its conditions.

Figure 7 shows the plant factory management GUI of the implemented mobile application.

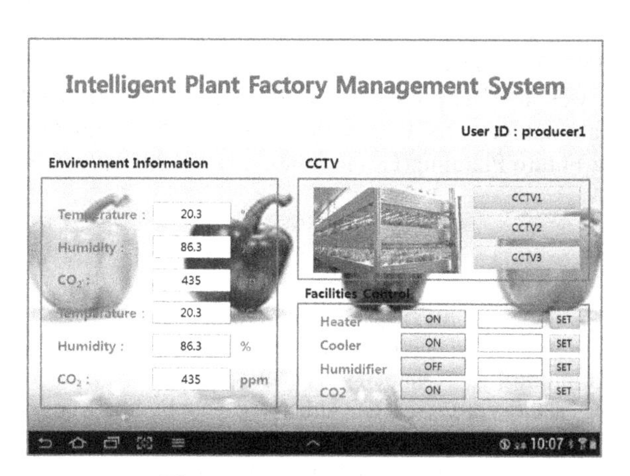

Fig. 8. Mobile application GUI

The mobile application is operated by JDK 1.6 in Window XP Service Pack3 OS, based on the basic tool of Eclipse 3.6 (Helios) for operation on Android. Android OS is the version of Android SDK 4.0 (Ice Cream Sandwich). It performs the identical functions to the web application. In any case of plant-factory abnormality, the application uses the notification function of Android to alert with sound or vibration via a mobile phone upon a real-time message receipt.

4 Conclusions

In this research, we proposed a ubiquitous computing-based intelligent plant factory system that collects contextual information including environmental conditions and nutrient solution necessary for crop growth through sensors, infers contextual information through the ontology and provides diverse services appropriate for each different situation through various application platforms.

For the proposed system, sensors need to be installed to measure environmental factors affecting intra-factory cultivation environment such as temperature, humidity, illumination intensity, CO_2, EC and pH. And based on the measured data, the system infers the contextual status of crop growth environment and stages through the ontology. Also, under the service rules defining the scope of available services for each situation as well as the inferred contextual information, the proposed system searches the optimal service for each given situation then triggers an event appropriate for the inferred service to automatically control the actuator for light, ventilation, humidity, CO2, concentration, temperature, etc. in line with its inference.

For the performance evaluation of the proposed system, we established ontology for intelligent control and intelligent plant factory. As a result of applying the developed system to the plant factory, we could find satisfying outcomes.

Acknowledgements. "This research was supported by the MSIP(Ministry of Science, ICT and Future Planning), Korea, under the CITRC(Convergence Information Technology Research Center) support program (NIPA-2013-H0401-13-2008) supervised by the NIPA(National IT Industry Promotion Agency)".

"This research was also supported by the International Research & Development Program of the National Research Foundation of Korea (NRF) funded by the Ministry of Science, ICT & Future Planning(Grant number: K 2012057499)".

References

1. UN world food programme (2012), `http://ko.wfp.org`
2. Park, S.C.: Green Agriculture and IT Fusion Cases of Korea Plant Factory. Journal of Korean Institute of Information Technology 9(1), 79–80 (2011)
3. Son, J.E.: Plant Factory - a Prospective Urban Agriculture. The Korean Society for Bio-Environment Control 2(1), 69–76 (1993)
4. Kim, J.H.: Trends and Prospects of Plant Factory. KREI Research report, vol. 61. Korea Rural Economic Institute, Seoul (2006)

5. Jeon, I.S., Min, J.H.: World Agriculture. KREI Research report: M45-118, Korea Rural Economic Institute, Seoul, Korea (2010)
6. Takamasamoto: Status of the Development of Complete Controlled Plant Factory. Electronic Engneering 8, 64–68 (2008)
7. Kim, Y.H.: Plant Factory Automation Systems and Technology Trends. KISTEP a Reserch Paper, Korea Institute of S&T Evaluation and Planning 5, 77–95 (2010)
8. Kim, J.W.: Trend and direction for plant factory system. Journal of Plant Biotechnology 37(4), 442–455 (2010)
9. Kim, Y.J., Han, H.S.: KREI agricultural administration Focus, vol. 49. Korea Rural Economic Institute, Seoul (2013)
10. Zato, C., De Paz, J.F., de Luis, A., Bajo, J., Corchado, J.M.: Model for assigning roles automatically in egovernment virtual organizations. Expert Systems with Applications 39, 10389–10401 (2012)
11. Pinzón, C.I., De Paz, J.F., Herrero, Á., Corchado, E., Bajo, J., Corchado, J.M.: idMAS-SQL: intrusion detection based on MAS to detect and block SQL injection through data mining. Information Sciences 231, 15–31 (2013)
12. Pinzón, C.I., Bajo, J., De Paz, J.F., Corchado, J.M.: S-MAS: An adaptive hierarchical distributed multi-agent architecture for blocking malicious SOAP messages within Web Services environments. Expert Systems with Applications 38, 5486–5499 (2011)
13. Protégé, `http://protege.stanford.edu`
14. OWL, `http://www.w3.org/TR/owl-features`
15. JENA, A Semantic Web Framework for Java, `http://jena.sourceforge.net`

Getting and Keeping Aged People Socially Included

Trials with Real End-Users of the EasyReach System

Michele Cornacchia, Filomena Papa, Enrico Nicolò, and Bartolomeo Sapio

Fondazione Ugo Bordoni, Viale del Policlinico 147, 00161 Rome, Italy
{mcornacchia,fpapa,nic,bsapio}@fub.it

Abstract. EasyReach is an Ambient Assisted Living Joint Programme project aimed at fostering the social interaction of home-bound and less educated elderly people. The project proposed a web interfaced service, accessed through the EasyReach DTT Set Top Box and a special remote control, equipped with interactive multimedia and an inertial unit. Pilots were carried out in the Italian cities of Rome and Milan by engaging real end-users in the controlled development of tasks. In the pilot conducted in Rome and described in this paper, data were collected by using both group and personal interviews, video recording and non-participant observations. Scenario engagement with realistic tasks was properly arranged.

In general, results show that EasyReach turns out to benefit people who stay at home or live alone, especially disabled people and those who are sick. The need to stay in touch is a primary driver for most, if not all, of the questioned elderly people. Retirement from active working life is a turning point in their lives and the feeling of being left at the margins of society often appears with different implications.

EasyReach is accepted as a way of spending the day and meeting friends, as well as to contact people one would not encounter otherwise, to join groups having the same interests and actively participating in social life. Sending messages can help sharing information and interests. Yet elderly people want more, asking to bridge the gap between them and public administrations, pension systems, healthcare, social services, emergency providers and similar.

Keywords: acceptance, adoption, aged people, ambient assisted living, behavioral change, digital divide, digital environment, e-inclusion, elderly, human computer interaction, human factors, innovation, isolation, resistance to change, scenario engagement, social inclusion, supported interaction, usage, user needs.

1 Introduction

The focus on Information and Communication Technologies characterizing most of the current policy actions on the information society in the European Union often fails to capture a serious challenge: e-inclusion is a moving target [1] if we consider that against a growth of ICT penetration in all groups of society, the digital divide has remained unvaried since the late 1990s. E-inclusion aims at ensuring nobody is left behind in the use of ICTs and all can enjoy the benefits of being socially included by

C. Ramos et al. (eds.), *Ambient Intelligence - Software and Applications*,
Advances in Intelligent Systems and Computing 291,
DOI: 10.1007/978-3-319-07596-9_11,

participating in the growing knowledge society. However, there are barriers to break down, especially affecting older people accessing services online [2], i.e. the lack of awareness, fear of using computer, fear of Internet safety, lack of user friendly interfaces, appropriate education, financial resources, lack of skills and understanding of how to use ICTs and online services (e-literacy). The digital divide in Europe is essentially age-related. Younger senior citizens (65-80 years old) are *de facto* those without access to digital and information technology: in 2011 two thirds of Europeans aged 65-74 have never used Internet [3]. Italy figures with 23 million non-users.

In order to effectively decrease the digital divide, the need for ICT-based services from the user's perspective must drive the realization of applications and services offered by the current technology. Services, communication and contents providers, industries and SMEs, in particular, must create an efficient market in which they can deliver in socially inclusive environments. In particular for the broadcast and broadband, this means hybrid or connected TV which brings together the worlds of broadcasting and the Internet on one single device. It is an ongoing convergence process, which has been described and discussed for over decades, and which will still evolve [4]. Manufacturers are trying to meet the consumer's demand for quality, interactivity [5] and personalisation of use. Gestures and voice control are recognised by many current users as potentially enhancing the connected TV experience, but the public are yet to really see the benefits [6]. Approaches to adapt interfaces at the specific needs and requirements are currently very limited [7], as elderly people still face problems when using digital devices in general. For them interaction is almost challenging, although accessible ICT applications could make difference for their living quality.

The Ambient Assisted Living Joint Programme (AAL JP) is scoring significant success in helping to create such favorable market conditions, since all projects must include a series of tests carried out by the final users [8]. Users are always involved in developing their own solutions. Accordingly to the AAL Call 2 track, the EasyReach project [9] proposes an ICT-DTT integrated based solution to elderly people for the advancement of e-inclusion [10] and social interaction. It mainly addresses the primary end-users, aged individuals [11, 12] with poor education and/or low income. The project assesses by qualitative measurements to what extent the proposed solution encounters a basic set of user needs and how the offer of the digital environment is attractive. The interface design is based on the traditional TV model of interaction, with its well-known and non-intimidating modalities including the remote control.

2 System and Social Environment

The idea at the basis of the project was to provide an ICT solution not requiring medium-high cognitive effort to be learnt. Given that aged people are familiar with TV sets and remote controls, EasyReach proposed a system including:

- a TV set (46 or 32 inch LCD);
- the EasyReach Remote Controller integrating: (i) a pointing service endowed with an inertial controller, (ii) a Hi-Res camera, and (iii) an audio recorder;

- the EasyReach Set-Top-Box (STB) allowing the user to access the system functionalities and join the EasyReach Network;
- an EasyReach Cloud, core of EasyReach Network, responsible to convey the system services and to manage all user information stored into a centralized database.

The STB runs the remote controller and the EasyReach Client, containing the software for (i) the social interaction environment, (ii) the personal assistant and (iii) the services required for a flexible connection to the system centralized database [13]. The EasyReach Social Interaction Environment aims at providing the elderly and pre-digital divide people with a simple and clean environment to interact with, where the information is easy-to-find and presented in a structured way. A sketchy view of the EasyReach user interface is shown in Fig. 1.

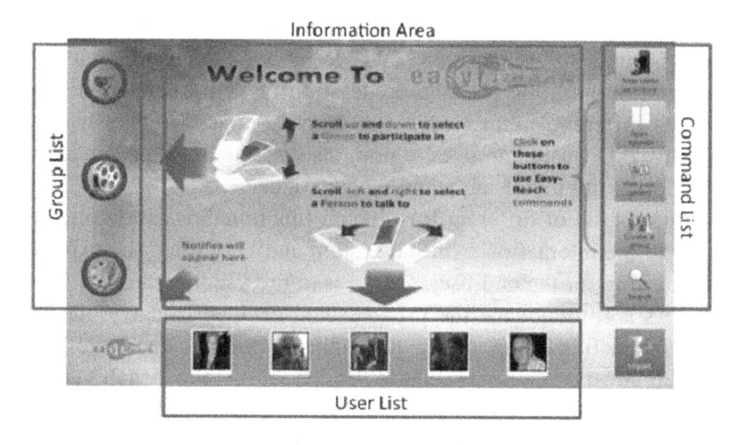

Fig. 1. The Social Interaction Environment

Two main sections can be identified on the screen:

1. **The Frame**, representing the static part of the environment and composed by the gesture driven scrollable contact bar (at bottom), containing the user's friends and relatives, as well as the people suggested by the Personal Assistant, the gesture-driven scrollable group bar (on left), where user can find his personal groups and the suggested ones, and a static command bar (on right), displaying the list of the actions available to the user (e.g. take a picture/video, create a group and so on).
2. **The Information Area**, representing the dynamic part that changes on the basis of the user's actions or selections. For instance, if the user selects a friend or a group, the Information Area will show the messages exchanged between the user and the selected element; similarly, if the user visits his gallery or wants to create a group, the Information Area will switch according to the user's selected action.

The remote control gesture recognition mechanism allows the user to navigate the system. Lists of users and groups can be scrolled by simple gestures, i.e. moving the controller right/left or up/down [13]. After selecting a given group or a friend, the user can exchange messages and/or updates within the related context. This function is made available by the system Social Channel which also draws easy-links between organizations and official institutions for larger information interchange. The making user's own expertise available to the whole community is of great value to maintain the self-perception of being socially active. In addition, the user can either create a new discussion group or join one suggested by the virtual Personal Assistant.

3 Trials Setting and Assessment Methodology

The scope of the field trials was to assess the social environment. Two pilots were planned to take place in different contexts, both geographical and cultural. They were carried out in the Italian cities of Rome and Milan by engaging real end-users in the controlled development of tasks. This paper presents the pilot in Rome.

Two Senior Centers ("Torrevecchia" and "Rebibbia - Ponte Mammolo") were selected in cooperation with the National Federation of Pensioners (FNP) to be representative of two different cultural areas in Rome. Some members having relevant roles in FNP of CISL, an Italian trade union, also attended as "privileged witnesses" of the elderly world with their long-standing experience.

The preliminary tests of different EasyReach functionalities soon proved a crucial lack of fluidity in the interaction with the system and very often the experience came out clearly frustrating for the end-user. Gestures to browse the interface were not very easy to learn. As a consequence the project team decided to carry out a "mediated experience of interaction", supporting the subjects with a "facilitator" role (room assistant) running most of the live scenario sequences on their behalf. This methodology was called "scenario engagement", aimed at providing an active setting for end-users in spite of the limitations with directly interacting with I/O devices. It was therefore possible to assess some components of the acceptance [14], the basic construct of the interaction model here applied, but related to a hypothetical use of EasyReach in the daily life.

The end-users sample was preliminarily identified in a set of 62 older adults (37 males, 25 females) with a mean age of 68.65 (SD = 5.88). Finally 40 subjects (21 males, 19 females) were recruited as participants to the pilot (30 individuals from the two Senior Centers - 15 for each Center - plus 10 others selected from FNP-CISL having also the role of "privileged witnesses"). Assessment of user acceptance was thus achieved by setting up overall 4 group sessions, by involving about 10 users in a single date.

As required by the AAL programme, demographic data of participants show a balanced distribution of genders (52% males and 48% females) with a peak in the age range 66-70 (32% of the total). Most of them were low educated (80% in the range "Primary Education" - "Secondary Education"). Almost all participants did use neither computers (85%) nor the Internet (72%).

Each trial session developed in about 80 minutes through the following main steps:

- demonstration of the main features of the system (remote control, user interface);
- "scenario engagement" of subjects in realistic tasks, i.e. social activities planning;

- collection of users' opinions of the acceptance aspects by interviews administered both to groups and privileged witnesses: perception of usefulness for social interaction and for improvement of the quality of life, willingness to use the system in the future, willingness to buy, facilitating conditions, etc.

As for the tasks, a couple of two simple scenarios were identified as the most fitting the usual life habits [15] of those people attending the two Roman Senior Centers joining the experimentation. They revealed to be very realistic:

Scenario 1 - "*A day to the cin*ema", theme aimed at organizing a group activity;
Scenario 2 - "Becoming a grandmother", theme aimed at sharing news with friends.

At the end of sessions, a little extra time too was dedicated to make EasyReach handy to anyone who was willing to try. Simple actions like recording audio messages, taking pictures or short video clips by using the remote control were made available.

4 Results

The full set of results derives from qualitative data analysis. Processed data consist of structured interviews from both group interviews and personal interviews to privileged witnesses, external research team observations and audio/video recordings. Subjects largely poured into the presented points of discussion the own personal experience of the old age in real world, in many cases gained by years of attendance at Senior Centers. Interviews were programmed in order to deeper explore the thoughts and the sentiments of participants just after the trials, by giving them time to think, no pressure and friendly dialogue atmosphere. Inevitably they also dealt with issues such as daily difficulties, the very spread condition of loneliness, and the bare necessities to which, until now, poor solutions were given by the existing social policies. The research team was able to assess some components of the acceptance and to group basic reaction patterns respectively on sentiments, opinions, desires, needs, interaction and behavioral resistance in relation with a conjectural use of EasyReach in the daily life.

Acceptance
Undoubtedly many people valued EasyReach as useful. Some of them showed enthusiastic views as well. In general, they said it can be useful especially for people who stay at home and live alone, for disabled and sick people. In particular, retirement from an active working life is a turning point in many lives and the fear of being left at the margins of society often appeared under different aspects.

4.1 Reaction Patterns

EasyReach could provide a practical response to basic communication needs, being rather easy to learn and clearly capable to change the daily lifestyle. Herewith a collection of synoptic thoughts about.

Sentiments
The system stimulates the elderly curiosity as well as it opens the horizon to several possibilities which can help the daily life. From the distance, it appears even not so complex and the environment provided is accepted as useful, especially to be connected to the outdoor world. EasyReach is seen as a possibility to win the sense of isolation, as it potentially promotes both the sentiment of sociality and the desire of shared culture.

Opinions
Opinions converge to few clear points addressing the improvement of functionalities and overall performance in view of possible indoor use. First of all the contact with the Public Administration, currently very complex and problematic to them, could be introduced but simplified through the EasyReach channels, especially to help those people with illness or handicap. Secondly, the necessity to improve the health care is pointed out as one of the basic lines to guide the future service developments. Third, in matter of personal security, many cases are reported about a subtle and persistent fear, a reason of constant anxiety due to some responsible social co-factors, not last the loneliness. Staying constantly in touch, sharing the same interests, participating more actively to the local social life, might produce the effect to create a great sense of serenity. At last, as for the willingness to pay for such a service, aged people have only one shared and reasonable opinion: "*yes, if it improves our quality of life*".

Desires
The most important desire stems from the acknowledgment that in many cases the elderly are not given a chance. They don't feel or think they were given a clear possibility, for example, to learn about new ICT devices, to update their knowledge, to lower the digital divide, to be of some help to the society by making their huge experiences available. They consider themselves not as fool or stupid people, only they have different basic needs and times of reaction. They are incline to promote and activate reciprocal assistance, to appreciate the possibility to help and receive help in case of need, to be constantly connected to their relatives, and as well as to give suggestions to improve the quality of social service delivering.

Needs
Sometimes elderly are not very acquainted with facts and problems of the outdoor nearby social life. It is not a fault to ascribe to anyone. Their resources are very often strongly taken to satisfy elementary needs, their minds diverted elsewhere, especially for those who fall in the lowest income brackets. EasyReach could really help these people by giving simple answers to simple needs, either material or social. For example, EasyReach could be configured as a tool for use by a group, thus including all, those familiar and not familiar to technology, and introducing a sort of good practice to share culture and useful information.

Interaction
Most of the time aged people need new skills to use ICT devices. It is a matter of fact that they see technology like a barrier too high to be overcome. This basic requirement has originally inspired EasyReach to be as easy as possible, however not all functions developed have perfectly matched the end-users skills and mental models. The "digital" complexity of the structure could not be smoothed or totally lowered. Crucial elements in the design of the user system interaction showed at times inadequate. Sometimes the same end-users specified ways to improve or remove these faults. The remote control is at the top, being out of an effective user control, then gestures follow as they are very challenging to be done.

Behavioral Resistance
Resistances to change can be assigned in general to every class of people independently from the age. As for the behaviour resistance to the adoption of EasyReach by aged people, this is subject to a narrow set of factors mostly ascribed to the difficulty of modifying personal habits if the proposed innovation doesn't really convince or respond to basic needs. Unfortunately EasyReach, besides many positive aspects, also raised this important sentiment. A relevant part of the subjects were not fully convinced of the potential advantages introduced by the system. They strongly intended to remain tied to their habits, even though they did not hide the desire to explore new horizons, to know more about new ICT devices, to constantly remain in touch with relatives, to do something innovative which can make them happy.

5 Conclusions

End-users' trials with EasyReach turned out realistic social digital experiences by aged people. Persons who attended were able to figure their own single ideas of use, scenarios' goals, service prototype advantages and limits, and be aware of the many benefits achievable online. They all participated with a lot of curiosity, showing great interest and willingness to contribute. The acceptance of the social environment given by the system was successfully evaluated, in spite of a lack of fluidity in operating the remote control with gestures. The need to stay in touch is a primary driver for most of elderly people. The most important barrier for usage is probably the critical mass needed to make the system useful: the ability to reach people also using web interfaces is considered to be crucial by most of the participants.

EasyReach is therefore considered an advantageous system, because it is a way of spending the day and sharing information, to be used to contact people one wouldn't encounter otherwise or to reach many others at the same time. The communication facilities are indispensable, yet elderly people want more, asking EasyReach to bridge the gap between them and Public Administration, pension systems, healthcare, social service, emergency providers and so on.

The colorful variety of different statements about intention to use reflects the spectrum of different attitudes towards life in general that elderly people have. A classification is complicated and somehow redundant, yet the micro-universe of the group interviews includes technology enthusiasts, total rejecters, seriously or superficially curious people, snobbish persons, money savvy customers. People who seriously need to be quickly included in the information society at conditions which they can accept.

References

1. Kaplan, D. (ed.): e-Inclusion: New challenges and policy recommendations. eEurope Advisory Group – WG2 – e-Inclusion: Final Report (July 2005)
2. Hardill, I.: E-Government and Older People in Ireland North and South. CARDI, Research commissioned by Office of First Minister and Deputy First Minister (October 2013)
3. European Commission, Digital Agenda Scoreboard 2012, Directorate-General for Communication Networks, Content and Technology (June 2012)
4. European Information Technology Observatory (EITO), Special report on The emergence of Connected TV services market (January 2012)
5. Quico, C., Geerts, D.: Interactive TV and video: Papers from EuroITV Conference, Introduction by Chairs. International Journal of Digital Television 3(2), 181–185 (2012)
6. Arch, A.: Web Accessibility for Older Users: A Literature Review. W3C Working Draft (May 14, 2008)
7. Jung, C., Hahn, V.: GUIDE - Adaptive User Interfaces for Accessible Hybrid TV Applications. In: A Position Paper for the Second W3C Web and TV Workshop, February 8-9 (2011)
8. `http://www.aal-europe.eu/about/objectives/`
9. `http://www.easyreach-project.eu/`
10. Guyader, H.L.: Einclusion public policies in Europe. Final report, EC (2009)
11. Rissola, G., Garrido, M.: Survey on eInclusion Actors in the EU27. In: Torrecillas, C., Centeno, C., Misuraca, G. (eds.) EC JRC, Final Draft (2013)
12. Senior Project, Ethics of e-Inclusion of older people, Discussion paper for the Workshop on Ethics and e-Inclusion, Bled (Slo) (May 12, 2008)
13. Bisiani, R., et al.: Fostering Social Interaction of Home-Bound Elderly People: The *Easy-Reach* System. In: Ali, M., Bosse, T., Hindriks, K.V., Hoogendoorn, M., Jonker, C.M., Treur, J. (eds.) IEA/AIE 2013. LNCS, vol. 7906, pp. 33–42. Springer, Heidelberg (2013)
14. Venkatesh, V., Morris, M.G., Davis, G.B., Davis, F.D.: User Acceptance of Information Technology: Toward a Unified View. MIS Quarterly 27(3), 425–478 (2003)
15. Sapio, B., Turk, T., Cornacchia, M., Papa, F., Nicolò, E., Livi, S.: Building scenarios of digital television adoption: a pilot study. Technology Analysis & Strategic Management 22(1), 43–63 (2010)

A General-Purpose mHealth System Relying on Knowledge Acquisition through Artificial Intelligence*

Giovanna Sannino[1,2], Ivanoe De Falco[1], and Giuseppe De Pietro[1]

[1] Institute of High Performance Computing and Networking, ICAR-CNR
[2] University of Naples Parthenope, Department of Technology
Naples, Italy
{giovanna.sannino,ivanoe.defalco,
giuseppe.depietro}@na.icar.cnr.it

Abstract. Remote monitoring of patients' vital parameters and ensuring mobility of both patient and doctor can greatly profit from real-time tele-monitoring technology. Here a description is given of a multi-purpose and multi-parametric tele-monitoring system. It can take advantage of the extraction, carried out offline and automatically on a desktop, of knowledge from databases containing measurements of patient's parameters. This knowledge is represented under the form of a set of IF...THEN rules that are provided to a rule-based mobile Decision Support System embedded in the system here presented. Then, wearable sensors collect in real time patient's vital parameters that are sent to a mobile device, where they are processed in real time by an app. If, as a consequence of the measured parameters, one of the above rules is activated, an alarm is automatically generated by the system for a well-timed medical intervention. Moreover all the monitored parameters are stored in EDF files for possible further analysis. This paper presents two practical applications of the system to two significant healthcare issues, i.e. apnea monitoring and fall detection. For these use cases, comparison with other well-known classifiers is carried out to evaluate the quality of the extracted knowledge.

Keywords: Wireless mHealth system, mobile monitoring, automatic knowledge extraction, IF...THEN rules, mobile DSS.

1 Introduction

m-Health applications aiming at detecting and monitoring anomalous conditions in real time are very important to improve citizens' health conditions and to reduce mortality and healthcare costs. In fact, aiming at saving time and at reducing costs, usually patients do not remain in hospital as a full recovery. Rather, they leave hospital as soon as possible and afterwards are periodically visited.

* This work has been partly supported by the project "Sistema avanzato per l'interpretazione e la condivisione della conoscenza in ambito sanitario A.S.K. – Health" (PON01_00850).

C. Ramos et al. (eds.), *Ambient Intelligence - Software and Applications*,
Advances in Intelligent Systems and Computing 291,
DOI: 10.1007/978-3-319-07596-9_12,

Recently, related technologies have quickly improved, so portable multi-life-parameter monitors have become a major research topic [1].

The technological improvements in measurement and information transmission, as sensor networks and wireless LAN, provide people with new options for the monitoring of vital parameters through the use of wearable sensors. Moreover, they allow patients to move without limitation while still being under continuous monitoring, so they help in achieving a higher quality in patients' care.

In this regard, our system is a multi-parametric and multi-purpose tele-monitoring system which allows having a global view about the health condition of patients being monitored, and it can be easily configured to detect different potentially dangerous situations. Our approach consists of two steps. The first step consists in *data analysis* and knowledge extraction carried out offline on a desktop starting from a database. Depending on the problem faced, this database contains either data from one single patient, leading to personalized knowledge, or data from a group of patients. The second step consists in the *real-time monitoring* through a suitable set of wearable sensors connected to a mobile device. An app on the mobile device processes the data coming from the sensors by exploiting the acquired knowledge. In this paper its use in two cases, i.e. apnea detection and fall detection, will be shown.

The main advantages of our developed system are: 1) the use of a friendly and adaptive user interface for multi-purpose monitoring; 2) its ability to provide real-time alert and alarm services; 3) no connection is required with any server because the reasoning about the gathered data is embedded on a mobile device with consequent battery saving; 4) the use of a Decision Support System in which the formalized experts' knowledge is embedded for an intelligent real-time monitoring; 5) if no experts' knowledge is available, the decision about whether or not a dangerous situation for the patient is taking place is based on knowledge automatically extracted offline from a database under the form of an explicit set of IF…THEN rules.

A consequence of the above issue 5) is that, depending on the available data, in some cases one set of rules can be extracted for each patient, thus leading to personalized healthcare.

Moreover, about the battery saving mentioned in issue 3), this comes from the fact that during real-time monitoring the mobile device has to manage one Bluetooth connection only with the sensors, rather than managing at the same time two wireless connections, namely a Bluetooth with the sensors and a wifi/3G towards a server or towards a Cloud service.

An analysis of the state of the art in literature shows that other systems recently developed for apnea monitoring [2-5] and fall detection [6-9] lack some important features presented by ours, and have also some other limitations which do not affect our system. For example, as far as apnea monitoring is concerned, although several other very recent proposals as e.g. [2], use mobile devices, these latter just gather data and send them to a hospital server where remote analysis is performed. The systems recently proposed in [3][4][5], instead, lack an action phase allowing performing immediate actions. Moreover, they all detect apneas based on general knowledge, unlike our system that is able to provide each patient with a set of rules specific for her/him.

As regards fall detection, instead, one of the most important limitations in other recent approaches is the lack of explicit knowledge formalized as a set of IF…THEN rules, which is very easily understood, user-friendly, and able to express well the reasons for a given decision. In fact, papers as [6] and [7] do not provide users with any explanation about their way of classifying, i.e, they are black boxes. For example in [6]a k-nearest neighbour algorithm is used to discriminate falls.

Besides, in other cases, e.g. [8] and [9] the rules are not extracted, rather they are generated by experts who examine fall recordings, guess the higher importance of some parameters with respect to others, and try to build some rules containing those parameters and the values of some thresholds. As an example, in [8] "a fall event is detected when acceleration magnitude is greater than 3 g and its peak is followed by a period, lasting at least 1200 milliseconds, characterized by the absence of peaks greater than the threshold". Of course, this approach is tiring and only a very limited number of parameters and of their combinations can be considered.

Section 2 describes the architecture of our mobile monitoring system. Section 3 presents two use cases, i.e. apnea monitoring and fall detection. Namely, the quality of the achieved results is discussed with reference to that provided by other classifiers. Finally, Section 4 contains our conclusions and future works.

2 Mobile Monitoring System Architecture

The system is based on a multi-layer architecture, as shown in Fig.1, to guarantee to it a set of properties as for example flexibility. Currently it is composed by three different layers, and each of them contains some modules. This type of architecture is very advantageous because, due to the independence between the layers, it allows easily adding algorithms and/or sensors as new modules in one layer without having knowledge about the upper one and the lower one. In the following we report on only some details, interested readers can refer to [10] and [11].

The Data Layer contains modules to manage sensors and patient data. In this layer data coming from the sensors are collected, and some vital parameters are calculated. In addition, it also contains modules for storing information in the European Data Format (EDF), the de-facto standard for vital signs recordings, and for saving monitored information in a personal Electronic Health Record (EHR) if desired.

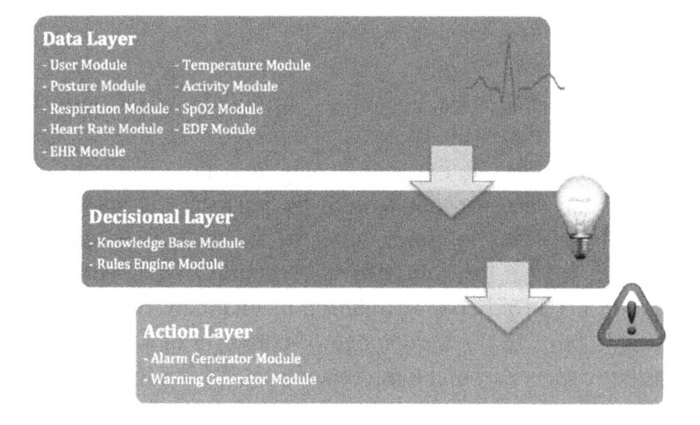

Fig. 1. Mobile Monitoring System Architecture

The Decisional Layer represents the intelligent core of the system and contains the rule engine detailed in [12]. It elaborates in real time data coming from the Data Layer. The Knowledge Base Module is composed by the personalized set of rules representing the formalization of knowledge about the monitored disease. Thanks to the Rule Engine module, the system recognizes in real time critical situations and determines the most suitable actions to be performed by the Action Layer.

The Action Layer executes the actions determined by the previous layer through the implementation of mechanisms that produce reactions as the generation of alarms.

To deal with the use cases described in next section, just one wearable sensor could be used, in which an ECG sensor and a 3-axis accelerometer are embedded. For example it's possible to use just a Zephyr Bio-Harness BH3 (www.zephyranywhere.com), an advanced Bluetooth physiological Monitoring sensor. Of course more sensors could be added, as for instance one for SpO2, so as to gather as many data as possible.

The system is implemented for mobile devices, as PDA and smartphone. It is developed by using Java, yet the system could be used to build desktop applications too, apart from the user interfaces. Namely, for these use cases the system was developed by using Eclipse IDE and Android SDK. A screenshot of the app is shown in Fig. 2.

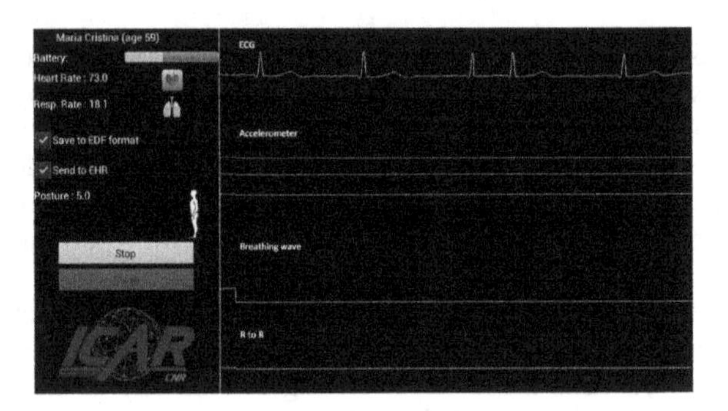

Fig. 2. A screenshot of the user interface

3 Use-Cases

Usually medical knowledge can be formalized in collaboration with domain experts. However, there are some cases in which is not possible to follow this methodology, therefore the knowledge must be extracted analyzing data available in medical databases. To deal with such cases, in our system we perform the automatic extraction of explicit sets of IF-THEN rules by using our DEREx tool [13]. This tool is based on Differential Evolution [14], a fast and effective version of an evolutionary algorithm. DEREx makes use of a 10-fold cross-validation mechanism for the offline selection of the set of rules that maximize the percentage of correctly classified items over unseen examples. Interested readers can make reference to [13] for details about DEREx.

3.1 Monitoring of Apneas

To face Apnea pathology the apnea-ECG database [15] has been taken into account. Starting from this database containing 70 patient recordings, we have created one for each of them in order to extract personalized sets of rules. Details about database creation, rule extraction and related results in terms of accuracy, sensitivity, and specificity can be found in [16].

To evaluate the goodness of the results in terms of accuracy, sensitivity, and specificity, other four well-known classifiers have been used for comparison. Therefore, the Waikato Environment for Knowledge Analysis (WEKA) system release 3.4 [17] has been used: it contains a large number of such techniques, divided into groups depending on the their basic working principles. From each such group a representative has been chosen. Among the Bayesian, the Bayes Net (BN) has been considered, and, among the function-based, the MultiLayer Perceptron Artificial Neural Network (MLP has been selected. The KStar (KS) has been considered as a representative of the lazy methods, while, among the tree-based, the J48 has been chosen.

It is important to state that none of these four classifiers outputs explicit sets of IF – THEN rules, rather each of them works as a black box and builds an internal model for the representation of the knowledge it acquires. Each such model is not user-friendly, i.e. it cannot be simply expressed/explained to a human being.

Also for them 25 runs have been performed, and 10-fold cross-validation has been carried out in each run. Table 1 shows the results in terms of the average accuracy (A) over the 35 databases, the standard deviation (StD), and the maximum (Max) and the minimum (Min) values.

Table 1. Results Achived by the Tested Classifiers in terms of Accuracy

	BN	*MLP*	*KS*	*J48*	*DEREx*
A	85.53	87.59	82.32	87.20	**92.26**
StD	9.40	8.62	11.37	8.83	**7.25**
Max	99.79	99.79	99.79	99.79	**100.00**
Min	65.67	70.82	65.82	72.10	**73.91**

For each parameter in Table 1, the best value obtained by all algorithms is shown in bold. The average percentage of correct classification provided by DEREX over the 35 databases is the highest, higher than the other ones by about 5%. Moreover, the standard deviation is the lowest, meaning that the algorithm is quite insensitive to the different initial random seeds. DEREx also achieves the highest maximum value, and is the only one to correctly classify 100% over five databases, i.e. those related to patients A01, A04, C05, C06, and C07. It also obtains the highest among the minimum values.

Table 2 contains the comparison of the algorithms in terms of average values of sensitivity over the 35 databases, and of the related standard deviation as well

Table 2. Average Results for Sensitivity.

	BN	*MLP*	*KS*	*J48*	*DEREx*
A	70.01	69.61	61.60	68.13	**82.14**
StD	35.58	35.60	34.61	36.23	**31.75**

The best values are reported in bold. DEREx achieves by far the highest value of sensitivity, so it has a much lower number of false negatives, i.e. apneas that are erroneously seen as non-apneas. This is very important for a system of this kind, as it is better able to generate alarms only when necessary.

Table 3 shows the same data for the specificity.

Table 3. Average Results for Specificity

	BN	*MLP*	*KS*	*J48*	*DEREx*
A	76.39	76.29	69.18	75.90	**79.40**
StD	**17.95**	20.23	25.00	20.22	31.26

Here too DEREx provides the highest value. This means that DEREx has a lower number of false positives, resulting in a lower number of unnecessary alarms. This comparison makes us confident that the set of rules shown is capable of discriminating apnea episodes much more accurately than these other artificial intelligence tools can.

Moreover, all these other classifiers do not extract rules, so they would be useless for the creation of the knowledge base needed by our system. We believe this user-friendliness of our system is very helpful to doctors.

3.2 Fall Detection

To deal with fall detection issue, we used the database built by Dr. S. Fudickar of the Dept. of Computer Science, Univ. of Potsdam, Germany [18]. Starting from this database containing 95 recordings, we have created one database from which extract knowledge as a set of rules. This has been done by means of three steps, i.e. annotation, windowing, and computation. Details about database creation, rule extraction and related results in terms of accuracy, sensitivity, and specificity can be found in [19].

To evaluate the goodness of the results in terms of percentage of correct classification, other four well-known classifiers have been used so as to compare DEREx against them. Also in this case, WEKA has been used. We have chosen a method based on Bayesian considerations, i.e. the Naïve Bayes (NB), a second based on functions and networks, i.e. a Radial Basis Function network (RBF), a third based on extraction of rules, i.e. OneR, and a fourth based on the occurrence of feature intervals per class, called Voting Feature Interval (VFI). For a fair comparison, they too have been run 25 times, as they too depend on an initial random seed, apart from Naive Bayes method, instead, which is deterministic, so it has been run only once.

Table 4 shows the results in terms of the average accuracy over those 25 values achieved, the related standard deviation, and the maximum and the minimum values among those 25. For each parameter in Table 4, the best value obtained by all algorithms is reported in bold.

Table 4. Results Achived by the Tested Classifiers in terms of Accuracy

	DEREx	***NB***	***RBF***	***OneR***	***VFI***
A	**91.88**	90,76	90,00	88,96	91,57
StD	0,53	---	**0,48**	1,54	0,78
Max	**92,92**	---	90,76	91,16	92,77
Min	**91,25**	---	89,16	86,35	90,36

The results in the table evidence that the average percentage of correct classification provided by DEREX over 25 runs is the highest, and so are the maximum and the minimum values achieved. Only for the standard deviation is RBF better, yet also in this case DEREx is very good, being the second best technique.

Table 5 contains the comparison of the five algorithms in terms of average values of sensitivity over the 25 runs, and of the related standard deviation as well. The highest average value and the lowest standard deviation are reported in bold. DEREx achieves by far the highest value of sensitivity, which means that it has a lower number of false negatives, i.e. falls that are erroneously seen as non-falls. This is very important for a system of this kind, as it is better able to generate alarms when necessary, i.e. when falls take place.

Table 5. Average Results for Sensitivity

	DEREx	***NB***	***RBF***	***OneR***	***VFI***
A	**91.3**	84.3	81.4	87.1	85.5
StD	0.23	---	**6e-5**	1e-3	4e-4

Table 6, instead, reports the same statistical values for specificity.

Table 6. Average Results for Specificity

	DEREx	***NB***	***RBF***	***OneR***	***VFI***
A	92.1	96.9	**98.1**	91.7	97.3
StD	0.21	---	3e-5	8e-4	**2e-5**

In this case RBF provides the highest value. The comparison between DEREX and RBF shows that our tool has better accuracy (by almost 2%) and better sensitivity (by almost 10%), whereas RBF is better in specificity (by exactly 6%), so DEREx is preferable. This comparison makes us confident that the set of rules shown is capable of discriminating falls more accurately than these artificial intelligence tools can.

Furthermore, most of these other classifiers do not perform rule extraction, so they would be useless for the creation of the knowledge base in our system.

Finally, the only other classifier capable of extracting rules, i.e. OneR, has average performance quite worse than that offered by our tool, since there is a difference of about 3% in accuracy, of 4% in sensitivity, and of 0.4% in specificity. This could be due to the working mechanism of OneR, that can generate sets of rules all containing one attribute only (the same for all the rules), as opposed to the flexibility in the number of attributes that can be contained in each set of rules proposed by DEREx.

4 Conclusion

The system presented here can be seen as a general-purpose one, as it can be used to easily deal with many different healthcare issues. It is based on the automatic extraction of explicit knowledge from databases by Differential Evolution, formalized under the form of a set of IF...THEN rules. Numerical results about tests on databases have shown the effectiveness of the approach, and the achieved sets of rules evidence its user-friendliness, which is in our opinion a very helpful feature to doctors.

Starting from the positive results in the two use cases, we plan to test our approach in cooperation with University Hospital ("Policlinico") of Naples in real-world situations: as concerns apneas, by means of a group of about twenty volunteering patients suffering from mild apneas, whereas, as regards falls, through a set of simulated falls performed in our iHealthLab.

References

1. Walker, B.A., Khandoker, A.H., Black, J.: Low cost ECG monitor for developing countries. In: 5th Int. Conf. on Intelligent Sensors, Sensor Networks and Information Processing, pp. 195–199 (2009)
2. Alqassim, S., Ganesh, M., Khoja, S., Zaidi, M., Aloul, F., Sagahyroon, A.: Sleep apnea monitoring using mobile phones. In: Proceedings of e-Health Networking, Applications and Services (Healtcom) 2012 Conference, pp. 443–446. IEEE Press (2012)
3. Wakchaure, S.L., Ghuge, G.D.: Apnea Pulse_Care: real-time data mining in sleep apnea monitor. Int. Journal of Emerging Technology and Advanced Engineering 3(4), 70–76 (2013)
4. Rofouei, M., Sinclair, M., Bittner, R., Blank, T., Saw, N., DeJean, G., Heffron, J.: A Non-invasive Wearable Neck-Cuff System for Real-Time Sleep Monitoring. In: Proc. of the 2011 Int. Conference on Body Sensor Networks, pp. 156–161. IEEE Computer Society, Washington, DC (2011)
5. Patil, D., Wadhai, V.M., Gujar, S., Surana, K., Devkate, P., Waghmare, S.: APNEA Detection on Smart Phone. International Journal of Computer Applications 59(7), 15–19 (2012)
6. Erdogan, S.Z., Bilgin, T.T.: A data mining approach for fall detection by using k-nearest neighbor algorithm on wireless sensor network data. IET Communications 6(18), 3281–3287 (2012)
7. Kerdegari, H., Samsudin, K., Rahli, A.R., Mokaram, S.: Evaluation of fall detection classification approaches. In: Proc. of ICIAS 2012, pp. 131–136 (2012)

8. Abate, S., Avvenuti, M., Cola, G., Corsini, P., Light, J., Vecchio, A.: Recognition of false alarms in fall detection systems. In: Proc. of 2011 IEEE Consumer Communications and Networking Conference (CCNC), pp. 23–28. IEEE (2011)
9. Sim, S.Y., Jeon, H.S., Chung, G.S., Kim, S.K., Kwon, S.J., Lee, W.K., Park, K.S.: Fall detection algorithm for the elderly using acceleration sensors on the shoes. In: Proce. of the 33rd Annual International Conference of the IEEE EMBS, pp. 4935–4938 (2011)
10. Sannino, G., De Pietro, G.: A Smart Context-Aware Mobile Monitoring System for Heart Patients. In: Proc. of IEEE BIBM, Atlanta, GA – USA, pp. 655–695 (2011)
11. Sannino, G., De Pietro, G.: An Evolved eHealth Monitoring System for a Nuclear Medicine Department. Developments in e-Systems Engineering (DeSE), 3–6 (December 6-8, 2011)
12. Minutolo, A., Esposito, M., De Pietro, G.: A Mobile Reasoning System for Supporting the Monitoring of Chronic Diseases. In: Nikita, K.S., Lin, J.C., Fotiadis, D.I., Arredondo Waldmeyer, M.-T. (eds.) MobiHealth 2011. LNICST, vol. 83, pp. 225–232. Springer, Heidelberg (2012)
13. De Falco, I.: Differential Evolution for automatic rule extraction from medical databases. Applied Soft Computing 13, 1265–1283 (2013)
14. Price, K., Storn, R., Lampinen, J.: Differential Evolution: A Practical Approach to Global Optimization. Springer (2005)
15. Penzel, T.: The Apnea–ECG database. Comput. Cardiol. 27, 255–258 (2000)
16. Sannino, G., De Falco, I., De Pietro, G.: Detecting Obstructive Sleep Apnea events in a realtime mobile monitoring system through automatically extracted sets of rules. In: Proc. of the 15th IEEE Int Conference on e-Health Networking, Application and Services (Healthcom 2013), Lisbon, Portugal (2013)
17. Hall, M., Frank, E., Holmes, G., Pfahringer, B., Reutemann, P., Witten, I.H.: The WEKA data mining software: an update. SIGKDD Explorations 11(1), 10–18 (2009)
18. Fudickar, S., Karth, C., Mahr, P., Schnor, B.: Fall-detection simulator for accelerometers with in-hardware preprocessing. In: Proc. of PETRA 2012, p. 41. ACM (2012)
19. Sannino, G., De Falco, I., De Pietro, G.: Effective supervised knowledge extraction for an mHealth system for fall detection. In: Proceedings of the XIII Mediterranean Conference on Medical and Biological Engineering and Computing (MEDICON 2013), Sevilla, Spain (2013)

Video Processing Architecture: A Solution for Endoscopic Procedures Results

Isabel Laranjo[1,2], Joel Braga[1,2], Domingos Assunção[1], Carla Rolanda[2,3,4], Luís Lopes[5], Jorge Correia-Pinto[2,3,6], and Victor Alves[1]

[1] CCTC - Computer Science and Technology Center, University of Minho, Braga, Portugal
[2] Life and Health Sciences Research Institute (ICVS), School of Health Sciences, University of Minho, Braga, Portugal
[3] ICVS/3B's - PT Government Associate Laboratory, Braga/Guimarães, Portugal
[4] Department of Gastroenterology, Hospital de Braga, Portugal
[5] Department of Gastroenterology, Santa Luzia Hospital, Viana do Castelo, Portugal
[6] Department of Pediatric Surgery, Hospital de Braga, Portugal
{isabel,valves}@di.uminho.pt,
{joeltelesbraga,dassuncao}@gmail.com,
{crolanda,jcp}@ecsaude.uminho.pt, luis.m.lopes@mac.com

Abstract. In this paper we propose an architecture for processing endoscopic procedures results. The goal is to create a complete system capable of processing any type of endoscopic multimedia results, in order to overcome the most common issues in the endoscopic domain (e.g. video's long-duration, gastroenterologist's possible difficulty to maintain the focus and efficiency during the viewing process, imperfections in images/videos). It was this scenario that led to the conception of the *MIVprocessing* solution, which will address these and other problems, providing an added value to the elaboration of diagnoses. The *MIVprocessing* is composed of five tasks: Video Summarization (elimination of the "non-informative" frames); Pre-Processing (correction/improvement of the frames); Pre-Detection; Segmentation; and Feature Extraction and Classification. The idea is to create a framework that brings together the capabilities of different but at the same time complementary concepts (e.g. image and signal processing, machine learning, computer vision). This conjugation applied to the endoscopic domain provides a set of features capable of improving the gastroenterologist's activities during and after the procedure.

Keywords: Endoscopy, White-light Endoscopy, e-Health, MyEndoscopy, MIVprocessing, Video Summarization, Image and Video Processing.

1 Introduction

Image processing techniques are used in several areas (e.g. Medicine, Archaeology, Biology). Within these areas, the computational tasks of image processing (e.g. contrast enhancement, adjusting intensity levels of gray, fix blurry images), are mainly used to improve the images visual perception by humans [1, 2].

C. Ramos et al. (eds.), *Ambient Intelligence - Software and Applications*,
Advances in Intelligent Systems and Computing 291,
DOI: 10.1007/978-3-319-07596-9_13,

Another area of application for image processing techniques is the resolution of problems related to the perception of the image by the computer. In this case, the goal is to obtain information contained in the image in a way that the computer can interpret it [2]. This information rarely has some sort of resemblance to the visual features that humans use for interpreting the image content. The coefficients of the Fourier transform, statistical information or multidimensional distance measures, are examples of processing techniques used for computational perception [2].

While humans (e.g. radiologist, gastroenterologist) can learn to interpret patterns in an image the computer solves the problem with processing techniques. Despite significant advances in image processing and in computer vision techniques, the nuances of knowledge extraction from an image still challenges the best algorithms [3].

Nowadays, digestive endoscopy is used to allow the gastroenterologist to locate two types of endoscopic findings (diseases and lesions), and to evaluate a wide variety of endoscopic diagnoses/diseases (e.g. esophageal varices, gastric and duodenal ulcer, benign and malignant tumors) or, in most cases, to ensure that the symptoms are caused by lesions (e.g. polyp - protruding lesions) [4]. The detection and classification of polyps is the prevailing domain of research for the development of computer-aided decision support system. Of all the approaches found in *Liedlgruber et al.* [5], 47 belong to polyps (in particular polyps located in the colon).

This paper is organized as follows: Section 2 describes the problem, in Section 3 we present the related work and in the next Section (4) we present the proposed solution (*MIVprocessing*). Finally, in Section 5 some conclusions are drawn and it is presented some future work.

2 Problem Definition

With the evolution of technology, the amount of information generated in the healthcare delivery has increased exponentially. In the endoscopic domain each procedure results in several gigabytes of multimedia information. This is related with the fact that sometimes the generated videos are too long. Depending on the type of endoscopic procedure, the videos may present a variable duration (e.g. Capsule Endoscopy $\approx$ 8 hours, Upper GastroIntestinal (GI) Endoscopy $\approx$ 25 minutes, Colonoscopy $\approx$ 40 minutes). The healthcare professionals can subsequently review the video in order to detect potential pathologies with a higher degree of confidence and accuracy, but at the same time, the time consuming reviewing process can make this task quite exhaustive and tedious.

Besides the time it takes to review the endoscopic procedure full-length video, this review is not always made in the best conditions due to technical issues, e.g. poor image quality, blurred image, the presence of specular reflection. These conditions make the presence of signs or symptoms that indicate the presence of pathology harder to detect.

These problems justify the conception of a solution for reducing the file size and duration of the video as well as improve the endoscopic images quality and the ability to assist the healthcare professionals on the detection/classification of endoscopic findings.

3 Related Work

From the capture and image processing systems found in the literature, there are some that stand as the basis for the several variants that have been developed. One of the better known base systems was proposed by *Gonzalez et al.* [2], which comprises a step for image/video acquiring, followed by the pre-processing, segmentation, feature representation and extraction, recognition and interpretation and, finally, the presentation of results. At each step, the information will be compared with a knowledge base.

Liedlgruber et al. presents, in its review work on endoscopic imaging [5], the main steps involved in computer-aided decision support system. Beyond the generic steps it includes a step for feature extraction and another for feature post processing. The classification step uses machine learning techniques.

Some recent studies have showed the potential of a Computer-Aided Diagnosis (CAD) system for the detection and classification of pathologies in endoscopic videos, providing the healthcare professional valuable technical support. There are CAD systems capable of detecting and/or classifying certain pathologies (e.g. polyps [6, 7], tumors [8], cancer [9], ulcers [6, 10], bleeding [11, 12]). Based on these studies a generic CAD system should follow a sequence of tasks: acquisition of image/video, low-level processing, segmentation and classification.

Other studies focuses on improving the quality of the image (e.g. elimination of specular reflection [13]) or detecting "non-informative" frames [14]. In these examples there are two common tasks: segmentation and classification. In 2007, *Lau et al.* [15] proposed a system for processing capsule endoscopic videos based on a framework that enables the analysis of multiple image features. This framework includes contrast enhancement tools, chamfer matching, threshold analysis, similarity between images, as well as color analysis tools.

All the studies referenced in this section implement some processing tasks that improve their own systems capabilities. However, there is no system integrating a complete set of tasks working together.

4 Proposed Architecture

The *MIVprocessing* was conceptualized from the idea of designing a processing system capable of grouping several tasks into a single framework, helping more effectively and accurately the healthcare professional at any time he is viewing, analyzing and/or interpreting the videos/images. The *MIVprocessing* is integrated in the *MyEndoscopy* system developed by *Laranjo et al.* [16]. The *MyEndoscopy* is a web-based application system, which allows the acquisition, processing, archiving and diffusion of endoscopic procedure results [16].

In Fig. 1 is presented a simple workflow describing the moments that occurs in a gastroenterology medical appointment in the healthcare institution that results in an endoscopy procedure (Moment 2). The *MIVprocessing* solution can be seamlessly integrated in the current workflow by performing some additional processing tasks, which can help the gastroenterologist's activities during and after the procedure.

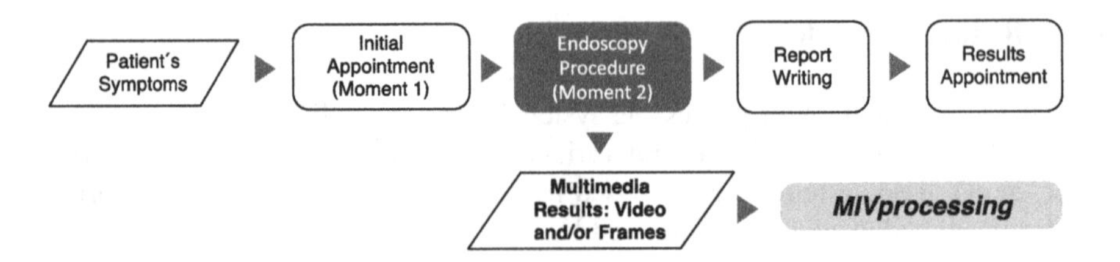

Fig. 1. Workflow of a gastroenterology medical appointment

The *MIVprocessing* results from the merger of two types of systems (**Supportive System** and **Computer-Aided Diagnosis (CAD) System**) and the tasks running on each of them (**Fig. 2**). The Supportive system consists of tasks that will support the CAD system. This means that these tasks are the basis of processing to be applied to videos, in order to make the detection and classification of abnormal pathologies more efficient and accurate in CAD system. The Supportive system is composed of three tasks (Video Summarization; Pre-Processing; and Pre-Detection), while the CAD system is composed of two tasks (Segmentation; Feature Extraction and Classification).

The *MIVprocessing* is integrated in *MIVbox*, which is a device for the acquisition, processing and storage of the endoscopic results.

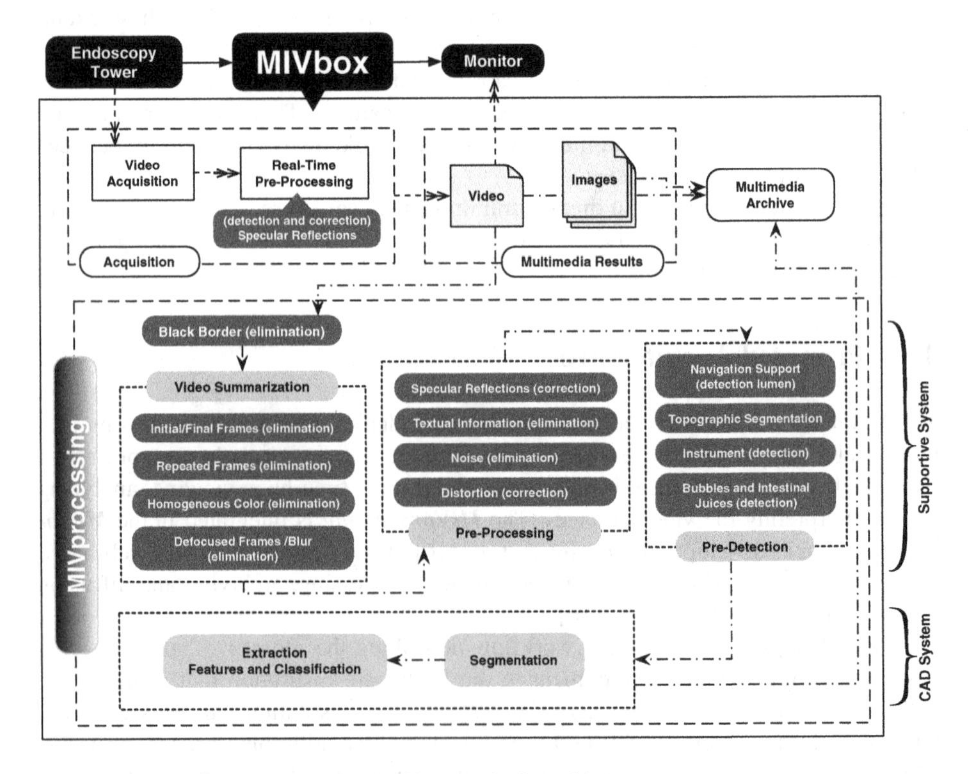

Fig. 2. MIVprocessing's architecture

Fig. 2 details the *MIVprocessing* architecture and also illustrates the steps that occur since video acquisition until archiving, not forgetting the visualization and processing steps. During the video acquisition and visualization (this process is based on the system described in [17]), *MIVprocessing* performs the detection and correction of specular reflections in real-time. The focus of this solution is in the white-light endoscopy imaging technique.

There is a preliminary task that deals with the elimination of the black borders, since these borders can interfere with the video processing, due to the contrast with the region of interest.

The main task, **video summarization**, aims to decrease the duration and file size of the full-length video, eliminating the "non-informative" frames (e.g. initial/final - captured outside the GI tract (Fig. 3.a), blurred (Fig. 3.b), homogeneous color (Fig. 3.c, Fig. 3.d), repeated) for easier visualization in the future by the gastroenterologist.

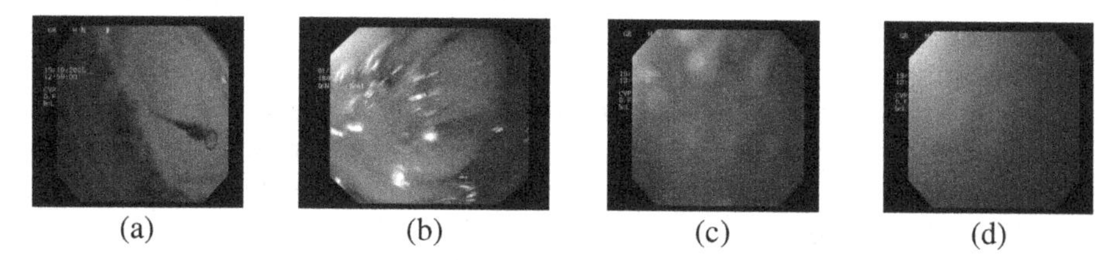

(a) (b) (c) (d)

Fig. 3. "Non-informative" frames: (a) initial, (b) defocused and (c, d) homogeneous color

Even after removing the frames considered not relevant, the reduced-length video still contains some information that, due to its characteristics, is considered unnecessary or even detrimental to the diagnosis. Some examples of this can be the noise present in the images, regions with textual information (Fig. 4.a) or regions with specular reflection (brightness) (Fig. 4.b). The goal of the main task, **pre-processing** (or low-level processing) is exactly to remove the information that is not necessary or even detrimental for diagnosis purposes. Some of the routines included are: cutting regions of interest, image rotation, scaling, color conversion, contrast enhancement or distortion correction.

The **pre-detection** is the last main task belonging to the Supportive system. Its routines are: navigation support, topographic segmentation, instrument detection ((Fig. 4.c), bubbles (Fig. 4.d) and intestinal juices detection and elimination. Detection of

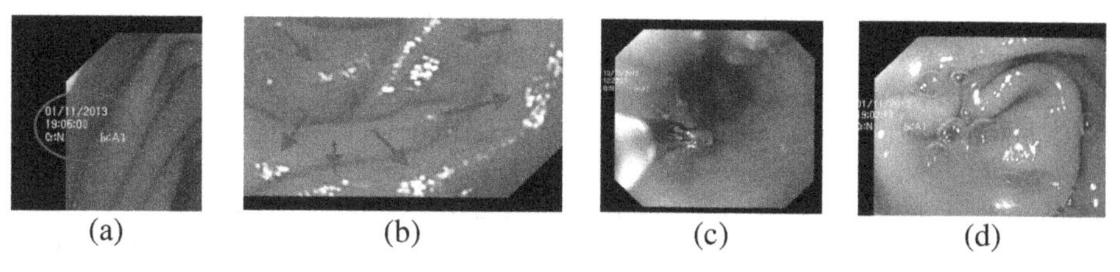

(a) (b) (c) (d)

Fig. 4. Regions with (a) textual information, (b) specular reflection, and frames with (c) instrument and (d) bubbles

the lumen (in navigation support) allows the extraction of important parameters for processing algorithms, e.g. determining the center and the border, which allows knowing the orientation of the GI tract. On the topographic segmentation occurs the video division into segments, specifically in different parts of the GI tract.

The purpose of the CAD system is to address the following issues:

- The correct detection of a given pathology;
- The exact location, along the GI tract, of the pathology found;
- Correct classification of the endoscopic findings.

The CAD system tasks aim to perform video segmentation, visual features extraction and its classification into semantic information. It also provides a reliable and accurate help at the time of diagnosis, never leading to the adulteration of the original information contained in the video. It is important to make clear that none of these tasks are intended to replace the gastroenterologist in the elaboration of diagnoses, since they are only support tools to help the diagnostic.

Segmentation is the task where the input data is divided into its constituents. The main role of segmentation is to distinguish the image background from the objects delimitation. The segmentation techniques used in this task are thresholding (Fig. 5.b), segmentation based on regions (Fig. 5.c), segmentation based on contours (Fig. 5.d) and classification by pixels.

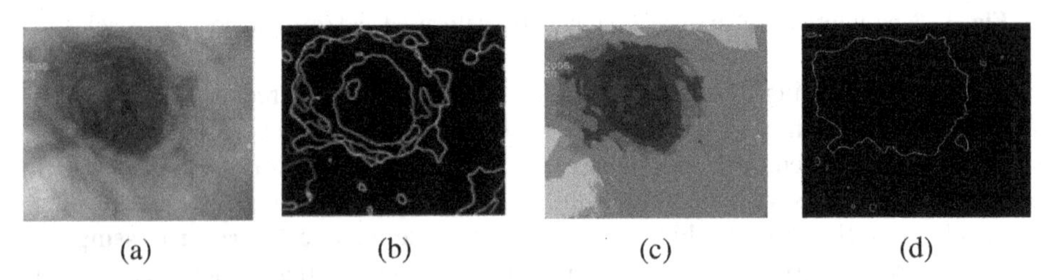

(a) (b) (c) (d)

Fig. 5. An example of image segmentation (a) original, (b) segmented - adaptive thresholding, (c) segmentation based on regions and (d) segmentation based on contours

At last, the CAD system has its **feature extraction and classification** task, which in a first moment selects and extracts the most important features (e.g. shape, texture, color) present in each frame and then, in a second moment, assigns a semantic designation to the endoscopic finding. The methodologies used in this task can be divided in two categories: Signal processing (processing information in the frequency domain, e.g. Fourier transforms, wavelet transforms) and Artificial intelligence (uses machine learning tools, e.g. Support Vector Machines (SVM), Bayes classifier).

As noted in [18], the clinical procedures and terminologies must be represented in expressive models that allow the storage of all relevant and necessary information. To ensure lexical compatibility of extracted and classified information, a standardized endoscopic vocabulary, the Minimal Standard Terminology (MST), is used [19]. The choice fell on the MST, in some part, because it is a terminology directed at Digestive

Endoscopy, but also for being complete and detailed (the terms described vary depending on the body part that is being examined and depending on the endoscopic technique), and easy to interpret (well organized and structured).

5 Conclusions and Future Work

In this paper, we gave an overview of the proposed solution (*MIVprocessing*). This is a solution that can be used with any type of endoscopic procedure – e.g. Capsule Endoscopy, Upper GI Endoscopy, Colonoscopy (with minor adjustments on the algorithms of feature extraction and semantic classification). In the case of Capsule Endoscopy, there is a slight difference once there is no detection step and elimination of specular reflection, in real-time.

MIVprocessing allows solving the problems described in section two with the following contributions:

- **Reduction in the duration and file size of the full-length video:** with the goal of creating a reduced-length video with only the informative frames;
- **Medical imaging quality improvement:** processing and/or elimination of detrimental information to the elaboration of the diagnosis;
- **Ability to assist the gastroenterologist in the detection/classification of endoscopic findings:** automatic detection of pathologies; defining the exact location of the detected pathology; and correctly classify the endoscopic finding.

The set of functionalities that are integrated in this solution makes it an asset to the quality increase in the provision of healthcare. Currently, the *MIVprocessing* is in the testing phase. Being integrated in *MyEndoscopy* it will provide this system with an extra set of functionalities, making it an optimal solution to the endoscopic domain.

From the studies referenced in the related work section, there is no study proposing an integrated system with a complete set of tasks working together. *MIVprocessing* has an architecture that integrates several pieces proposed in the literature and builds a complete endoscopic processing system.

Future work will be focused on fine tuning the proposed solution and evaluating it not only at the technical level, but also at the functional level by using end users like gastroenterologists and researchers. The assessment of the medical experts' opinion is also an important step towards optimizing the task of classification of endoscopic pathologies, since these experts are the ones that can identify the most relevant features for correct identification of pathologies.

Although the *MIVprocessing* has been designed for the white-light endoscopy imaging technique it can be extended and adapted to other techniques (e.g. Narrow Band Imaging (NBI), Confocal Laser Endomicroscopy (CLE)).

Acknowledgments. This work is funded by ERDF - European Regional Development Fund through the COMPETE Programme (operational programme for competitiveness) and by National Funds through the FCT - *Fundação para a Ciência e a Tecnologia* (Portuguese Foundation for Science and Technology) within project

FCOMP-01-0202-FEDER- 013853 and by National Funds through the FCT - *Fundação para a Ciência e a Tecnologia* (Portuguese Foundation for Science and Technology) within project PEst-OE/EEI/UI0752/2014.

References

1. Moeslund, T.B.: Introduction to Video and Image Processing: Building Real Systems and Applications. Springer, London (2012)
2. Gonzalez, R.C., Woods, R.E.: Digital Image Processing. Pearson Education International (2009)
3. Bui, A.A.T., Taira, R.K., Kangarloo, H.: Introduction - What is Medical Imaging Informatics? In: Bui, A.A.T., Taira, R.K. (eds.) Medical Imaging Informatics, pp. 3–14. Springer US (2010)
4. Schiller, K.F.R., Warren, B.F., Hunt, R.H.: Atlas of Gastrointestinal Endoscopy and Related Pathology. Wiley-Blackwell (2002)
5. Liedlgruber, M., Uhl, A.: Computer-Aided Decision Support Systems for Endoscopy in the Gastrointestinal Tract: a Review. IEEE Rev. Biomed. Eng. 4, 73–88 (2011)
6. Karargyris, A., Bourbakis, N.: Detection of Small Bowel Polyps and Ulcers in Wireless Capsule Endoscopy Videos. IEEE Trans. Biomed. Eng. 58, 2777–2786 (2011)
7. Alexandre, L.A., Casteleiro, J.M., Nobreinst, N.: Polyp Detection in Endoscopic Video Using SVMs. In: Kok, J.N., Koronacki, J., Lopez de Mantaras, R., Matwin, S., Mladenič, D., Skowron, A. (eds.) PKDD 2007. LNCS (LNAI), vol. 4702, pp. 358–365. Springer, Heidelberg (2007)
8. Barbosa, D.C., Roupar, D.B., Ramos, J.C., Tavares, A.C., Lima, C.S.: Automatic Small Bowel Tumor Diagnosis by using Multi-scale Wavelet-based Analysis in Wireless Capsule Endoscopy Images. Biomed. Eng. Online. 11, 1–17 (2012)
9. Iakovidis, D.K., Maroulis, D.E., Karkanis, S.: a: An Intelligent System for Automatic Detection of Gastrointestinal Adenomas in Video Endoscopy. Comput. Biol. Med. 36, 1084–1103 (2006)
10. Chen, Y., Lee, J.: Ulcer Detection in Wireless Capsule Endoscopy Video. In: Proceedings of the 20th ACM International Conference on Multimedia, pp. 1181–1184. ACM (2012)
11. Pan, G., Yan, G., Qiu, X., Cui, J.: Bleeding Detection in Wireless Capsule Endoscopy Based on Probabilistic Neural Network. J. Med. Syst. 35, 1477–1484 (2011)
12. Li, B., Meng, M.Q.-H.: Computer-Aided Detection of Bleeding Regions for Capsule Endoscopy Images. IEEE Trans. Biomed. Eng. 56, 1032–1039 (2009)
13. Stehle, T.: Removal of Specular Reflections in Endoscopic Images Removal of Specular Reflections in Endoscopic Images. Acta Polytech. J. Adv. Eng. 46, 32–36 (2006)
14. Bashar, M.K., Kitasaka, T., Suenaga, Y., Mekada, Y., Mori, K.: Automatic Detection of Informative Frames from Wireless Capsule Endoscopy Images. Med. Image Anal. 14, 449–470 (2010)
15. Lau, P.Y., Correia, P.L.: Analyzing Gastrointestinal Tissue Images using Multiple Features. In: Proceedings of the International Conference on Telecommunications, pp. 435–438 (2007)
16. Laranjo, I., Braga, J., Assunção, D., Silva, A., Rolanda, C., Lopes, L., Correia-Pinto, J., Alves, V.: Web-Based Solution for Acquisition, Processing, Archiving and Diffusion of Endoscopy Studies. In: Omatu, S., Neves, J., Rodriguez, J.M.C., Paz Santana, J.F., Gonzalez, S.R. (eds.) Distrib. Computing & Artificial Intelligence. AISC, vol. 217, pp. 317–324. Springer, Heidelberg (2013)

17. Braga, J., Laranjo, I., Assunção, D., Rolanda, C., Lopes, L., Correia-Pinto, J., Alves, V.: Endoscopic Imaging Results: Web based Solution with Video Diffusion. Procedia Technol. 9, 1123–1131 (2013)
18. Oliveira, T., Novais, P., Neves, J.: Guideline Formalization and Knowledge Representation for Clinical Decision Support. Adv. Distrib. Comput. Artif. Intell. J. 1, 1–12 (2012)
19. Aabakken, L., Rembacken, B., LeMoine, O., Kuznetsov, K., Rey, J.-F., Rösch, T., Eisen, G., Cotton, P., Fujino, M.: Minimal Standard Terminology for Gastrointestinal Endoscopy (MST 3.0) (2009)

A Multi-agent Platform for Hospital Interoperability

Luciana Cardoso[1], Fernando Marins[1], Filipe Portela[2], Manuel Santos[2], António Abelha[1], and José Machado[1,*]

[1] University of Minho, Computer Science and Technology Centre, Braga, Portugal
[2] University of Minho, Algoritmi Research Centre, Guimarães, Portugal
{a55524,a55561}@alunos.uminho.pt,
{cfp,mfs}@dsi.uminho.pt,
{abelha,jmac}@di.uminho.pt

Abstract. The interoperability among the Health Information Systems is a natural demand nowadays. The Agency for Integration, Diffusion and Archive of Medical Information (AIDA) is a Multi-Agent System (MAS) specifically developed to guarantee interoperability in health organizations.

This paper presents the Biomedical Multi-agent Platform for Interoperability (BMaPI) integrated in AIDA and it is used by all hospital services which communicates with AIDA, one of the examples is the Intensive Care Unit. The BMaPI main objective is to facilitate the communication among the agents of a MAS. It also assists the interaction between humans and agents through an interface that allows the administrators to create new agents easily and to monitor their activities in real time. Due to the BMaPI characteristics it is possible ensure the continuous work of the AIDA agents associated to INTCare system.

The BMaPI was installed in *Centro Hospitalar do Porto* successfully, increasing the functionality and overall usability of AIDA platform.

Keywords: Hospital Interoperability, Multi-Agent Systems, Agents Monitoring, AIDA, INTCare.

1 Introduction

Implementing technology in health organizations is increasing exponentially. The Health Information Systems (HIS) are one of the biggest examples of these technologies, which have played a key role in the workflow in these organizations nowadays. However, these systems are distributed and heterogeneous. The interaction among these systems is a crucial demand these days. In this way, the interoperability among the HIS becomes an indispensable feature in health organizations [1, 2].

Interoperability is the capacity of two systems interact between them, ensuring the understanding of the process and data exchanged on both sides [3].

* Corresponding author.

C. Ramos et al. (eds.), *Ambient Intelligence - Software and Applications*,
Advances in Intelligent Systems and Computing 291,
DOI: 10.1007/978-3-319-07596-9_14,

Standards such as the Health Level Seven (HL7) avoid different structures of the transferred information and in this way the correct interpretation and communication are achieved among the HIS [4]. There are various technologies that have the ability to implement the interoperability among these systems, such as the Service Oriented Architecture (SOA), webservice interfaces, eXtended Markup Language (XML) and Multi-Agent Systems (MAS) [5]. This last one has demonstrated to be a powerful technology in the area of interoperability, addressing heterogeneous and distributed limitations [1,5,6].

The Agency for Integration, Diffusion and Archive of Medical Information (AIDA) is an agent-based platform with the purpose of ensuring the interoperability among HIS [6]. INTCare is an Intelligent Decision Support System to Intensive Care Units which is also based on a MAS and it uses some data provided by AIDA platform to construct their models [7]. The correct working of AIDA agents it is fundamental to the success of INTCare system. However it fails in controlling and monitoring its own agents.

In this context, it emerges the Biomedical Multi-agent Platform for Interoperability (BMaPI). It was integrated into the AIDA platform and it enables the AIDA administrators to verify the agents functioning or to detect eventual failures in their performance in real time. Through the BMaPI, the administrators can manage successfully all the AIDA agents, knowing with the detail when a specific agent performs its activities. So, the administrators are able to select the most opportune period to execute changes, updates, maintenance and other operations, improving the AIDA performance.

This paper is divided into five sections. This first section introduces the work presented in this paper and the contextualization is also presented. The second section is a background that aims the importance of agents and MAS for the interoperability and it also presents the AIDA platform. Section 3 presents the archetype for monitoring the AIDA agents: the BMaPI, more specifically its features, its architecture and the description of its components. The results obtained after BMaPI implementation are demonstrated and discussed in Section 4. The last section exposes the major conclusions and the future work.

2 Background

2.1 Multi-Agent Systems for Interoperability

The multi-agent technology has excelled in the interoperability implementation in the HIS [1,6,7]. This technology is rather close to the concepts that characterize a distributed architecture. The success of the agent-based computing has strongly appeared as a result of its ability to work out problems and to make a new revolution in the development and analysis of software [6].

Intelligent agents are computational artifacts that are endowed with some proprieties such as the autonomy, reactivity, pro-activity and social skills. The autonomy and pro-activity are features that allow to plan and to perform tasks designed to achieve the proposed objectives, without direct human intervention. The reactivity enables the agents act according to the environment where they are inserted. Through their social skills, the agents interact with each other in order to obtain a common purpose. With these properties, the MAS demonstrate to be a strong technology for the interoperability implementation among the HIS [1,6,7].

2.2 AIDA

Techniques based on Artificial Intelligence (AI) have shown great potential when introduced into the hospital environment. Most of these systems are focused in the area of systems integration and decision support systems [8]. So, the Agency for Integration, Diffusion and Archive of Medical Information (AIDA) is a solution developed by a research group of AI at the University of Minho. AIDA is already implemented in several portuguese health organizations, including the *Centro Hospitalar do Porto* (CHP). It is based on the agent-oriented paradigm and its MAS increases according to the necessity of each institution. This platform has shown great adaptability, modularity and effectiveness in the health organizations which is implemented. AIDA was created with the intent to aid medical applications and to manage the information flow through processing systems with an adaptable level of autonomy. The main objective is to make the HIS interoperable and to provide complementary tactics of diagnostics and therapeutics, through the diffusion and the integration of the information produced in a health organization. The agents are the basic unit of AIDA and they ensure the communication among heterogeneous systems, sending, receiving, managing and storing information and responding to requests timely and correctly [5,6].

INTCare is an Intelligent Decision Support System optimized to Intensive Care Units and it is associated to AIDA. The main objective is provide new knowledge to the decision process automatically and in real-time [8]. INTCare platform uses intelligent agents [7] to support their process and automate some tasks. In this context, the BMaPI became an important platform not only to AIDA, but also to INTCare because it allows to monitor the agents which are directly associated to the system.

3 Archetype for Agents Monitoring

The Biomedical Multi-agent Platform for Interoperability was designed to allow MAS administrators to control and to manage a community of agents, allowing their survival and the interoperability in a heterogeneous environment. It is a platform that integrates all agents belonging to the MAS, regardless the machine where they perform their tasks. This is intended to ensure not only that all agents communicate each other, but also to create an interoperable environment. The communication among the agents are made through messages that meet certain structural rules defined by the FIPA (Foundation for Intelligent Physical Agents) called ACL (Agent Communication Language) [9].

According to the needs described by administrators of the AIDA platform, it was developed the BMaPI which aims: Ensure a greater control over the agents that constitute the AIDA; Facilitate the user's work in the creation and registration of new agents locally or remotely; Allow the user to enable and to disable services at the health unit, through the launch or stop of a particular agent; Facilitate the scheduling and rescheduling of the agents activity; and Monitor dynamically and in real time the agents activity.

3.1 Architecture and Components Description

The BMaPI was constructed respecting the specifications of a client/server architecture, where the server is able to communicate with multiple clients simultaneously dedicating a thread to each of them. The ACL messages are exchanged between the agents through sockets, which use TCP/IP protocols to ensure that information is transferred from an agent to another keeping the integrity of transferred data.

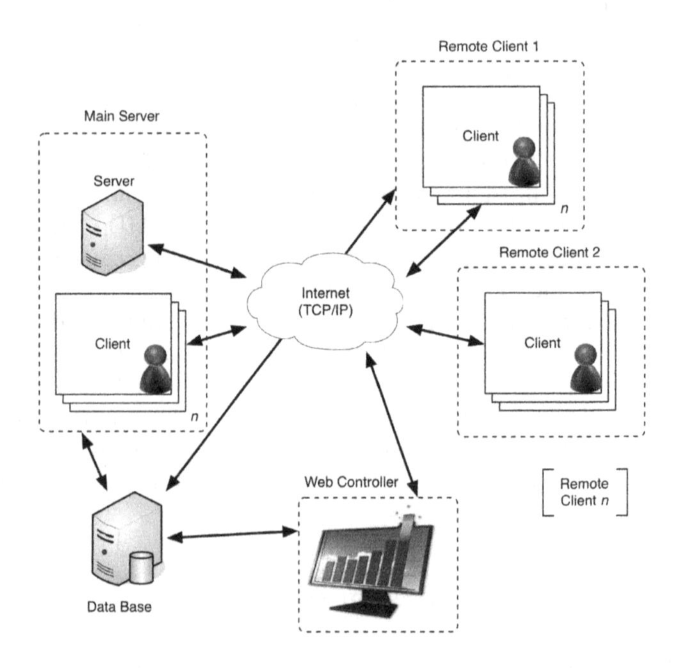

Fig. 1. Architecture of BMaPI

Analyzing the BMaPI architecture presented in the Figure 1, is possible verify that it is composed of three distinct components: the Main Server, the Remote Clients and the Web Controller. The first two components are similar, the first is a server for the entire system and the second is the main client, which is installed on each machine and its purpose is to connect to the Main Server. The Web Controller is a user interface to control the agents, which allows the user to schedule the activities of each agent and also to monitor their activities.

Main Server and Remote Client
The Main Server is the only mandatory component, it makes the automatic boot (Figure 2) of the whole platform and it can be executed only once. The machine where this component runs is the main machine because it serves as the server but it can also host agents. The Remote Clients can be executed simultaneously on multiple machines, in all those that host agents. The main difference between these two components is that the Main Server might work remotely, i.e it has the ability to create and to send the agents to perform their activity in other machines.

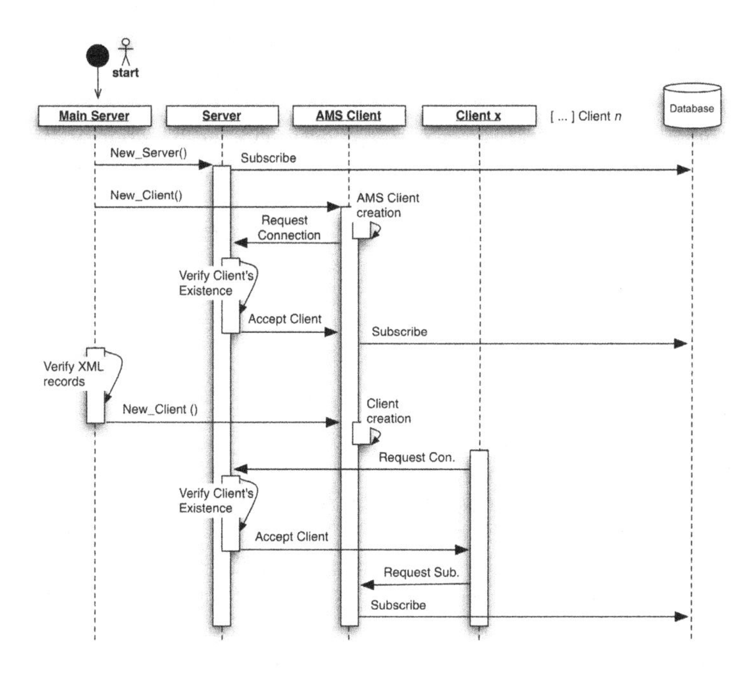

Fig. 2. Automatic boot process of BMaPI

Analyzing the Figure 2 that represents the initialization of the BMaPI, it is possible to verify that the server creation is the first step followed by a record in the database of the IP and the port where it was created. After this the server is waiting for new connections from new clients. The first client automatically created and connected to the server is the Agent Management System (AMS). It is a special agent that exists in each machine and it is responsible for controlling all other agents hosted on this machine. After the AMS acceptance by the server, to finalize the creation process, some AMS information is stored in the database, namely its name, its state (active or inactive) and the machine IP wherein it operates. So that the administrator does not have to integrate all the agents in BMaPI each time there is a startup (of the machine or of the BMaPI), the agent information is stored locally in an XML file. This file serves so that whenever the platform is initialized all agents are created automatically.

The Remote Clients are started like the Main Server with the exception of the creation of the server. In these cases the request connection by the agents is sent to the server situated in the main machine where is the Main Server component.

Web Controller
The Web Controller constitutes the interface with the humans and it provides an attractive environment to control and to monitor the agents subscribed on the BMaPI. Exploiting each agent, it is possible access three pages: to visualize its properties; to schedule its activities; and to monitor its performed activities in real time and in a dynamic way. The properties page enables not only to analyze the properties of the agents, but also a summary of all activities performed by the agent since its creation until the present moment. More precisely: the date of the last activity; the duration of the last activity; the number of times that the agent was executed; the average duration of all activities of the agent; and the number of errors that occurred. The scheduling page gives the administrator a set of possibilities to program the activity of each agent. And the monitoring page has two kinds of dynamical graphs that correspond to the number of times that a specific agent performed its tasks and the average duration of the activities of the agent.

4 Implementation Results

In order to test the BMaPI functionalities, it was implemented on the CHP a hospital in the North of Portugal. A set of AIDA agents was integrated in BMaPI and their activities were controlled. The results presented in this paper are related to the period between 10 and 16 of September 2013 and to the agent 609. This agent called 609 is responsible to ensure the interoperability with the support system to the practices of nursing and its results are selected to be presented due to it be an agent that generally has some problems in its activities.

These results (Figure 3) represent the average duration of the activity of the agent 609 in seconds clustered per hour on September 11, 2013. From the graph presented in the Figure 3, it is concluded that the agent took less time to execute its tasks between 7AM and 8AM, with an average of 281.83 seconds and the maximum value was detected between 11AM and 12PM, with a duration average of 457.62 seconds. It is possible to analyze that there is an abrupt growth between 8AM and 12PM followed by a gradual decrease until 11PM. This trend of the graph line can be related to the amount of data recorded in the nursing information system and consequently with the influx of patients to the nursing service. With all information that the web controller provides the administrators of the AIDA or of another MAS wherein the BMaPI was implemented can make a better manage of overall system and its subsystems (agents). They can know the period when the agents take more time to perform their activities and consequently decide which will be the best time to make routine changes to

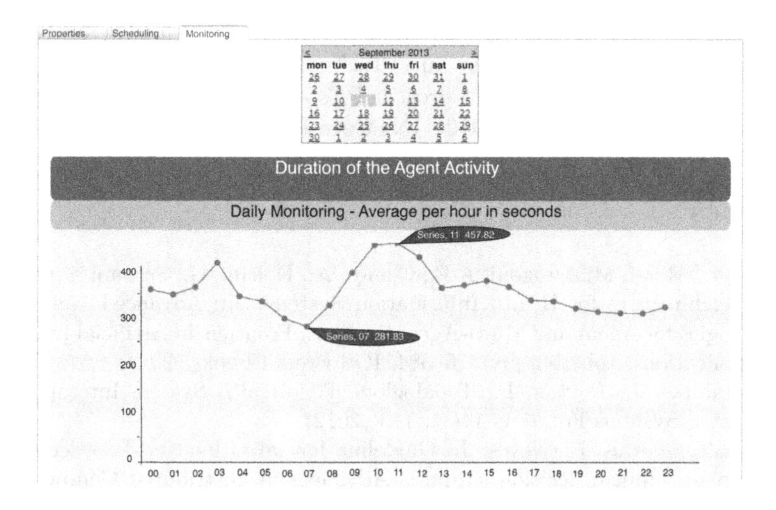

Fig. 3. Daily analysis of the duration of the activity of the agent 609 on September 11, 2013

system upgrades, and other operations to improve overall system performance. Therefore the BMaPI when applied in a MAS improves its functionality, once it promotes functions that satisfy the users' needs. And it also offers more usability to the MAS, since being a system of easy understanding and operation with an attractive interface. With the BMaPI the interoperability is assured because the agents are able to communicate with each other in order to achieve mutual goals.

5 Conclusions and Future Work

It is possible concluded that the BMaPI implementation in the CHP offers to its AIDA and INTCare System a greater control of their agents, improving the functionality and the usability of both systems. INTCare is one of the system which it was directly benefited with the new features of AIDA. The BMaPI components developed with the agent-oriented programming revealed to be a powerful tool to control minutely these agents and to avoid possible failures and consequently, to improve the quality of the services delivered and the interoperability among the HIS. Besides that, a specific MAS (for example the AIDA or the INTCare) administrators are able to know the best period to apply operations like maintenance, updates or other changes, minimizing the disturbances in the MAS workflow.

In the future, a communicative link can be established between BMaPI and a system developed by our research group, which enables the monitoring and the prevention of faults from the AIDA agents and machines [10]. This communication will enable to balance resources of AIDA (or other MAS) through the migration of its agents.

Acknowledgement. This work is financed with the support of the Portuguese Foundation for Science and Technology (FCT), with the grant SFRH/BD/70549/2010 and within project PEst-OE/EEI/UI0752/2014 and the contract PTDC/EEI-SII/1302/2012 - INTCare II.

References

1. Palazzo, L., Rossi, M., Dragoni, A.F., Claudi, A., Dolcini, G., Sernani, P.: A Multi-Agent Architecture for Health Information Systems. In: Advanced Methods and Technologies for Agent and Multi-Agent Systems. Frontiers in Artificial Intelligence and Applications, vol. 253, pp. 375–384. IOS Press Ebooks (2013)
2. Weber-Jahnke, J., Peyton, L., Topaloglou, T.: eHealth System Interoperability. Information Systems Frontiers 14(1), 1–3 (2012)
3. Oliveira, T., Novais, P., Neves, J.: Guideline formalization and knowledge representation for clinical decision support. Advances in Distributed Computing and Artificial Intelligence Journal (ADCAIJ) 1(2), 1–12 (2012)
4. Khan, W.A., Hussain, M., Latif, K., Afzal, M., Ahmad, F., Lee, S.: Process interoperability in healthcare systems with dynamic semantic web services. Computing 95(9), 837–862 (2013)
5. Miranda, M., Duarte, J., Abelha, A., Machado, J., Neves, J.: Interoperability in Healthcare. In: Proceedings of the 24th European Simulation and Modelling Conference (ESM), Hasselt, Belgium. Eurosis (2010)
6. Peixoto, H., Santos, M., Abelha, A., Machado, J.: Intelligence in Interoperability with AIDA. In: Chen, L., Felfernig, A., Liu, J., Raś, Z.W. (eds.) ISMIS 2012. LNCS, vol. 7661, pp. 264–273. Springer, Heidelberg (2012)
7. Santos, M.F., Portela, F., Vilas-Boas, M.: Intcare: multi-agent approach for real-time intelligent decision support in intensive medicine. In: Proceedings of the International Conference on Agents and Artificial Intelligence (ICAART 2011), pp. 364–369. SciTePress (2011)
8. Portela, F., Santos, M.F., Machado, J., Abelha, A., Silva, Á.: Pervasive and Intelligent Decision Support in Critical Health Care Using Ensembles. In: Bursa, M., Khuri, S., Renda, M.E. (eds.) ITBAM 2013. LNCS, vol. 8060, pp. 1–16. Springer, Heidelberg (2013)
9. Foundation for Intelligent Physical Agents (FIPA): FIPA ACL Message Structure Specification (2002), `http://www.fipa.org/specs/fipa00061/SC00061G.pdf` (accessed on February 22, 2014)
10. Marins, F., Cardoso, L., Portela, F., Santos, M., Abelha, A., Machado, J.: Improving High Availability and Reliability of Health Interoperability Systems. In: Rocha, Á., Correia, A.M., Tan, F., Stroetmann, K. (eds.) New Perspectives in Information Systems and Technologies, Volume 2. AISC, vol. 276, pp. 207–216. Springer, Heidelberg (2014)

Mobile Solution Using NFC and In-Air Hand Gestures for Advertising Applications

Francisco Manuel Borrego-Jaraba, Gonzalo Cerruela García, Nicolás García Pedrajas, Irene Luque Ruiz, and Miguel Ángel Góme-Nieto

University of Córdoba, Department of Computing and Numerical Analysis, Campus Universitario de Rabanales, E-14071 Córdoba, Spain
{fborrego,gcerruela,npedrajas,iluque,mangel}@uco.es

Abstract. In this paper we present an application based on Near Field Communication technology and oriented to advertising and user loyalty fields. The system combines NFC and in-air hand gestures with the purpose of creating an attractive leisure activity where customers are rewarded by shops with the possibility of obtaining a prize or discount on the purchases made. NFC is used for securing the participation of the user in the activity and for exchanging the prizes obtained. During the activity, users perform a fishing gesture (throwing and catching) with the mobile phone, simulating the capture of a prize that could be located anywhere in the shop. This gesture is recognized by mobile sensors. The tailoring of the system to specific business marketing strategies is also managed by the proposed solution.

1 Introduction

Near Field Communication technology [1, 2] is widely used in marketing and advertising applications. Most of these applications are based on the use of PoS (Point of Sale) as augmented objects with a tag providing the user with some information, discount vouchers or whatever other type of service needed with the aim of promoting a commercial product. In these applications, when users with a NFC enhanced device touch the PoS, they immediately receive the information related to the marketing campaign.

In the last years, several applications have been developed using NFC for marketing and advertising [3, 4] purposes. WingBonus [5] manages different types of discount vouchers (coupons, bonds and chits) that users can freely download from the Web portal or by simply touching a tag located in a shop or PoS. In this system, the vouchers' life cycle is fully managed (from downloaded the exchanged status). Security of vouchers is managed by means of synchronization processes with the server. Users can exchange the vouchers in the allowed shops by touching a NFC reader, a tag in the shop (or using QR technology). WingBonus also manages loyalty cards. Any number of loyalty cards can be managed by the mobile application while deposit and withdraw operations are performed in the same way as with vouchers.

Different applications have considered the useful application of NFC to advertising, marketing or user loyalty [6] purposes. Hence, objects or screens augmented with

C. Ramos et al. (eds.), *Ambient Intelligence - Software and Applications*,
Advances in Intelligent Systems and Computing 291,
DOI: 10.1007/978-3-319-07596-9_15, © Springer International Publishing Switzerland 2014

tags are used to provide users with information about products, sales or services. The secure element or SIM has been the main proposed way to support the credit or loyalty cards identification.

Besides this, the use of gestures in mobile computation has been generating a growing interest in recent years [7]. The idea is the development of interfaces close to the human behavior for the building of easy and more adaptable pervasive applications. Researches and developments in gestures cover two main areas: camera-based and movement sensor-based [8]. In camera-based applications, body or face user's movements are captured by cameras in order to determine the user's activity, interest, felling and so on. Thus, for instance, large screens used for advertising can show information to users when they are detected by a camera or the display can change the information when the camera detects a specific face expression in the user. Gesture applications based on movement sensors are mainly oriented to the management of touch screens. Different finger movements can be detected by the device's sensors in order to facilitate the communication with the applications. Other types of movement sensor applications are based in body movements. In these applications, measurements from the compass and accelerometer of the mobile device are used as input to calculate the movement performed by the user; for instance, walk, lift or turn the phone, etc. Using the information gathered from the mobile sensors (as well as user location) these applications can guide the user in a city or a museum, exchange files, connect different devices, or control them (TV, radio, lamps, etc.), perform user identifications, etc. [9].

In this paper, we describe a system called WingTrapper that combines both aforementioned technologies: in-air hand gestures and NFC. WingTrapper is an application oriented to advertising and user loyalty purposes; offering companies the capacity to distribute discount vouchers or any other kind of offers in an easy and amusing way.

In this system, shops provide to users participation tickets following their marketing strategies. Tickets can be distributed by means of RFID tag, NFC readers or even QR codes. By touching the tag, users gather the participation ticket that is managed by the mobile application. Afterwards, users can use this ticket by following its requirements in order to get the prizes of the marketing campaign established by the shop.

The use of a ticket is performed by means of gestures; a fishing gesture (throwing and catching) should be made by the user, simulating the capture of a prize located at any place in the shop. Finally, if the prize is obtained by the user, this prize is stored by the mobile application allowing its later exchange in the shop, again using NFC technology.

In this paper, we describe the architecture of the systems, and the characteristics of the application context. The mobile application is also described showing an indoor application scenario.

2 System Architecture

At first, WingTrapper was created as a new utility of an application oriented to the management of m-coupons and loyalty cards called WingBonus [9]. WingBonus

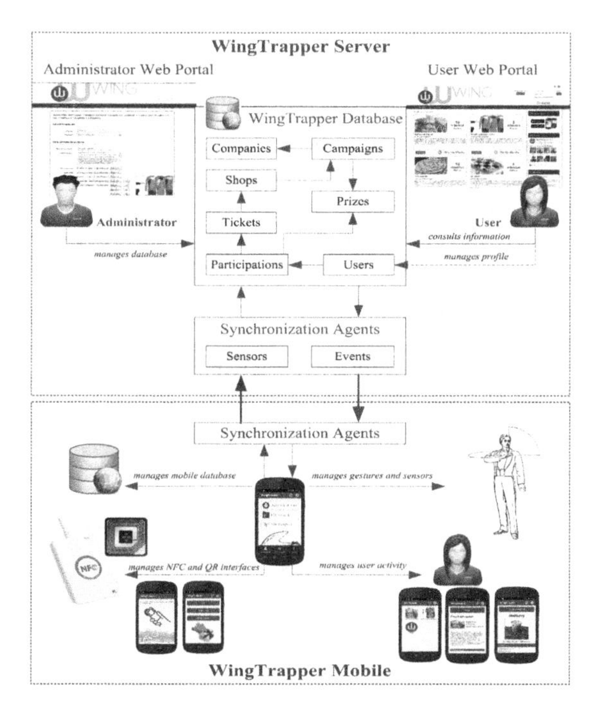

Fig. 1. WingTrapper architecture

manages any kind of m-coupons or vouchers allowing companies to promote their products thanks to the use of discount vouchers. Vouchers and loyalty cards are stored in the mobile phone; therefore, they are available to users at anytime without the risk of losing or forgetting them.

In WingBonus, users can obtain the vouchers in different ways: a) downloading them from the Web portal, b) touching a tag or reading a QR code located anywhere, and c) using NFC or QR technology from a PoS. WingBonus's mobile application allows the user to fully manage the vouchers and loyalty cards stored in the device. Marketing campaigns are managed in different folders by the mobile application and the users can consult the voucher details and exchange them in any of the allowed shops. Moreover, shops provided with NFC readers can gather information from the mobile application applying, without user intervention, the discounts that match the stored vouchers.

WingTrapper extends WingBonus functionality, although it can also be used as an independent application, allowing any shop or company to develop tailored marketing campaigns oriented to product advertising and user loyalty. WingTrapper is composed by two main components: the server and mobile subsystems.

As shown in Figure 1 the server is composed of different components. The administrator Web portal is in charge of storing the information corresponding with the marketing campaign. This component is used for the management of the information about companies, shops, participation tickets, users, marketing campaigns and prizes.

The user Web portal is only an advertising portal related to the active marketing campaigns. This portal shows information about the marketing campaigns and prizes to the users, allowing them to download the mobile application and register with it.

The server side includes a synchronization component in charge of synchronizing the mobile and server databases. All the information about users, tickets, participations, prizes, etc., is securely stored on both sides. User's activities are gathered by the mobile application and synchronized with the server side in order to ensure data reliability.

The mobile application is the core of WingTrapper. This application has been developed in Java. Users equipped with a NFC phone can gather participation tickets corresponding to marketing campaigns distributed by the shops. These tickets are stored in the mobile phone allowing the user to participate in the campaign and to try to obtain one or more of the prizes offered.

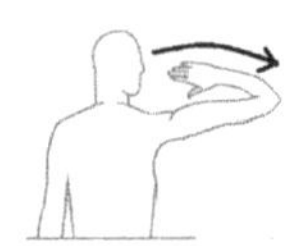
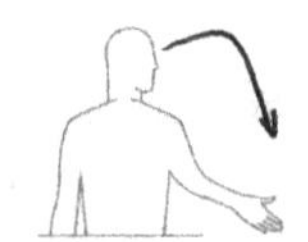
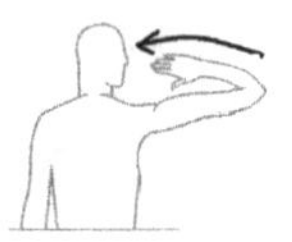

Fig. 2. Fishing gesture

The participation consists in trying to fish a prize, this prize could be located anywhere in the shop. Therefore, by performing a throwing and catching movement with the mobile phone, as shown in Figure 2, the user participates in trying to catch the prize. The user has as many attempts as allowed in the participation ticket. If a prize is captured, information about the prize is stored in the mobile application for their later exchange in the shop. Prizes can correspond to some product, discount or even vouchers or money for the loyalty card managed by WingBonus; in this case this information is also stored by the WingBonus application.

The Context in WingTrapper

WingTrapper is a system though as a different and funny activity for promoting advertising and user loyalty products. Customers are rewarded with the possibility to win some gift or bonus when they carry out some purchase in the shop. The system can be used in indoor and outdoor scenarios allowing a fully customization of the marketing business strategy. The main elements involved in the system are described below:

Marketing Campaigns

WingTrapper manages any type of marketing campaigns consisting in offering users any kind of prizes. For instance, when the user makes a purchase from a shop of a certain quantity, or a specific number of times or number of products, during the happy hour or day, etc., the shop offers the user a participation ticket to win some of the offered prizes.

Each marketing campaign is managed in a different folder, and it has textual and graphic information associated, allowing easy identification by the user.

Each campaign has an associated a validity date, information about the participation tickets, the prizes, the way the prizes are distributed throughout in the environment and the requirements for their capture. This information allows companies to tailor their marketing campaigns, even at each shop level.

The Prizes

Prizes are anything the company can offer users as a reward for its loyalty or purchase. Prizes are associated with campaigns being uniquely identified by the system, also storing the private identification used by the shop. WingTrapper stores textual and graphic information of prizes and other information oriented to the customization of the campaign to the marketing strategies, as follows:

- Assignation of specific prizes to different shops.
- Location of each prize in the shop.
- The distribution frequency of each type of prize: a) attempts by ticket, b) number of distributed tickets, c) attempts by user, d) characteristics of random distribution, e) characteristics of user loyalty, f) etc.

Prizes' location in the shop allows companies to improve their advertising strategy, moving customers to specifics areas or sale zones in order to improve their possibility of capturing some prize.

The Tickets

Participation tickets are distributed to the users depending on the company's marketing strategy. Tickets can be downloaded from any Web portal and gathered by the user from a NFC reader or tags located in PoS or anywhere.

Tickets store information about the campaign. Each ticket is uniquely identified and allows the user to participate in the prizes' capture. Each ticket has the associated number of permitted attempts, validity dates, allowed shops and, in order to customize the business strategy, tickets can have private information such as: whether the ticket has associated or not a prize or the number of attempts to get the prizes, among others.

Brief and detailed information about tickets is managed by the mobile application. User can consult their own tickets, the remaining participations, the prizes obtained, etc. This information is stored in the server, so users cannot lose it even if they delete it.

Moreover, ticket security is managed by the system in two ways: a) the ticket identification is double; the server side assigns a unique identifier to the ticket (always hidden to the user) and the own ticket identification managed by the shop, b) the use of the tickets can be restricted to a validation process (depending on the information defined in the marketing campaign). This process is carried out by the server and the mobile application and it may need the participation of the shop.

The Prize Capture

The capture of prizes is performed by means of a gesture, as shown in Figure 2. User performs a fishing gesture simulating the catching of an object that could located in some place of the surrounding environment. By doing a movement of throwing and catching with the mobile phone in the user's hand, WingTrapper detects the gesture, gathering the information from the accelerometer and compass mobile sensors, as well as the GPS user location (when needed).

Information of sensors, shop, user and ticket is sent to the server side for analysis and validation. Once the information is validated, the participation event is tested against the campaign characteristics stored in the database. Finally, the result is sent back to the user and if a prize has been obtained, its information is stored in the mobile phone.

3 A Real Application Scenario of Using the Mobile Solution

In order to show the characteristics of the system, we describe in this section one of the application scenarios where the system has been applied. We describe the characteristics of the environment and the marketing strategies proposed by the shop.

The Scenario and the Marketing Campaign

The indoors environment is a pub in the city of Córdoba. The pub rewards its customers with a participation ticket for each drink/course. The number of attempts included in the participation ticket depends on the value of the bill. Thus, tickets that range from one to five attempts have been generated. As many customers did not own a NFC enhanced phone, the participation tickets were cards including a NFC tag and a QR code. The prizes won by customers are also exchanged in the pub using cards with NFC tags and QR codes. Users' mobile phone should be connected by GPRS or WiFi at any moment.

The customer is told that the prizes are flyer bonds flying inside the pub, so he/she will try to fish one of these flyer bonds with a throwing and catching gesture performed with his/her mobile phone.

Different type of prizes were defined: a) discount vouchers from 5% to 10%, b) free drinks (wine, beer or soda), and c) WingBonus chits (from one to three chits), that is, a kind of voucher considered in WingBonus application, consisting in points to be exchanged in the pub for some service.

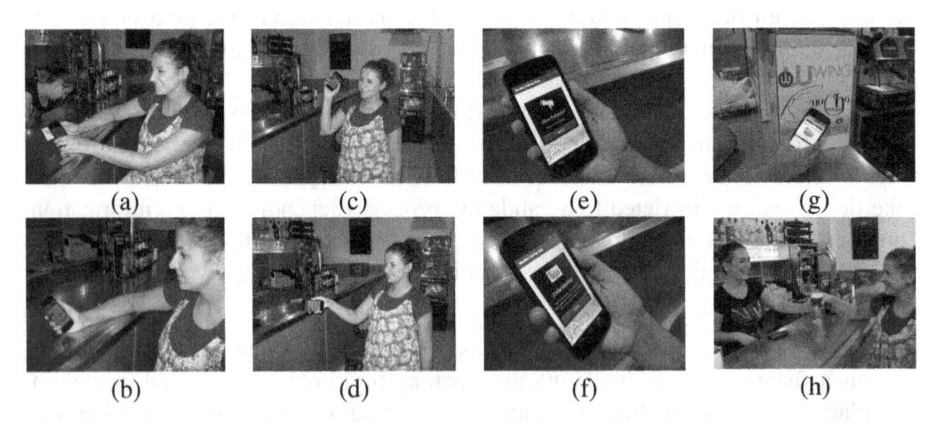

Fig. 3. Some snapshots of ticket dispatching

Showing the Activity

Figures 3 (a) to (h) show some images captured from the use of WingTrapper in this scenario. Figure 3 (a,b) shows how the participation ticket is given to the customer. The ticket is stored in the mobile phone and server databases and the user can review it.

Figure 3 (c,d,e,f) shows the user participation and the result obtained. The user tries to catch some *trappy* in a funny experience with a bad result. Finally, in the last attempt the user wins a trappy consisting in a free drink.

Finally, Figure 3 (g,h) shows some snapshots of the prize management by the mobile application and the exchange process of the prize in the pub. In this process, the user selects the prize in the application and he/she touches the exchanging tag. In this process the information is sent to the server and a "receipt" is sent back to the application in order to be confirmed by the pub manager.

4 Conclusion

In this paper we have described the combined use of Near Field Communication technology with in-air hand gestures recognition for the building of pervasive advertising and loyalty systems.

Near Field Communication technology is used for the management of the user participation in the event, the ticket validation and collection of the prizes obtained. The event consists in the capture of prizes disseminated in the surrounding environment. In order to capture a prize, the user should perform a fishing gesture with the mobile phone in his/her hand; simulating the attempt to catch a flyer trappy. Once the prize has been obtained, the user can exchange it using NFC technology again.

Information and prize security is managed by the server and the mobile phone databases, synchronization processes, and unique object identification exchanged in the NFC transactions.

Marketing campaigns are tailored according to company strategies. Thus, WingTrapper allows the customization of the number and type of prizes and its location, user's participation rules, rules of prizes assignation such as user loyalty, value of the user purchase, date, hour and user location, shop, etc.

The combined use of NFC and in-air hand gestures allow us to propose a fresh and funny system oriented to improving companies' strategy for products advertising and user loyalty; offering customers rewards of any type, at the same time as capturing the attention of other customers when watching people being entertained during the event.

Acknowledgment. The Ministry of Science and Innovation of Spain (MICINN) supported this work (Project: TIN2011-24312).

References

[1] NFC Forum (2013), http://www.nfc-forum.org/home

[2] Garrido, P.C., Miraz, G.M., Ruiz, I.L., Go, X., Mez-Nieto, M.A.: A Model for the Development of NFC Context-Awareness Applications on Internet of Things. In: Second International Workshop on Near Field Communication, pp. 9–14 (2010)

[3] Wiedmann, K.P., Reeh, M.O., Schumacher, H.: Near Field Communication im Mobile Marketing. In: Bauer, H., Bryant, M., Dirks, T. (eds.) Erfolgsfaktoren des Mobile Marketing, pp. 305–325. Springer, Heidelberg (2009)

[4] Holleis, P., Broll, G., Böhm, S.: Advertising with NFC. In: The Eighth International Conference on Pervasive Computing, pp. 1–10. Worldpress, Helsinki (2010)

[5] Sánchez-Silos, J.J., Velasco-Arjona, F.J., Ruiz, I.L., Gomez-Nieto, M.A.: An NFC-Based Solution for Discount and Loyalty Mobile Coupons. In: Proceedings of the 2012 4th International Workshop on Near Field Communication, pp. 45–50. IEEE Computer Society (2012)

[6] Riekki, J., Salminen, T., Alaka, X., Rppa, I.: Requesting Pervasive Services by Touching RFID Tags. IEEE Pervasive Computing 5, 40–46 (2006)

[7] Asadzadeh, P., Kulik, L., Tanin, E.: Gesture recognition using RFID technology. Personal Ubiquitous Comput. 16, 225–234 (2012)

[8] Milosevic, B., Farella, E., Benini, L.: Continuous gesture recognition for resource constrained smart objects. In: The Fourth International Conference on Mobile Ubiquitous Computing, Systems, Services and Technologies, UBICOMM 2010, pp. 391–396 (2010)

[9] Mayrhofer, R., Gellersen, H.: Shake Well Before Use: Intuitive and Secure Pairing of Mobile Devices. IEEE Transactions on Mobile Computing 8, 792–806 (2009)

Ubiquitous Sensorization for Multimodal Assessment of Driving Patterns

Fábio Silva, Cesar Analide, Celestino Gonçalves, and João Sarmento

University of Minho
Department of Informatics, Braga, Portugal
{fabiosilva,analide}@di.uminho.pt,
celestin@ipg.pt, joao.sarmento@gmail.com

Abstract. Sustainability issues and sustainable behaviours are becoming concerns of increasing significance in our society. In the case of transportation systems, it would be important to know the impact of a given driving behaviour over sustainability factors. This paper describes a system that integrates ubiquitous mobile sensors available on devices such as smartphones, intelligent wristbands and smartwatches, in order to determine and classify driving patterns and to assess driving efficiency and driver's moods. It first identifies the main attributes for contextual information, with relevance to driving analysis. Next, it describes how to obtain that information from ubiquitous mobile sensors, usually carried by drivers. Finally, it addresses the multimodal assessment process which produces the analysis of driving patterns and the classification of driving moods, promoting the identification of either regular or aggressive driving patterns, and the classification of mood types between aggressive and relaxed. Such an approach enables ubiquitous sensing of personal driving patterns across different vehicles, which can be used in sustainability frameworks, driving alerts and recommendation systems.

Keywords: Driving Profile, Mobile Sensors, Sustainability.

1 Introduction

Ambient Intelligence (AmI) is a very active area of knowledge and constitutes a multi-disciplinary subject which takes advantage of advances in sensing systems, pervasive devices, context recognition, and communications. Nowadays, AmI applications can be found in fields ranging from home, office, transport, tourism, recommender and safety systems, among many others [20]. In the case of transport applications, an area also known as Smart Cars [2], the AmI system must be aware not only of the car situation, but also of the driver's intention, of his physical and physiological conditions and of the best way to deal with them [19], [17]. The driver's behaviour is, thus, of key importance: several authors have used machine learning and dynamical graphical models for modelling and recognizing driver's behaviours [22].

There are examples of applications integrating AmI and ubiquitous principles in driving and traffic analysis. In [13], it is described a monitoring and analysis

C. Ramos et al. (eds.), *Ambient Intelligence - Software and Applications*,
Advances in Intelligent Systems and Computing 291,
DOI: 10.1007/978-3-319-07596-9_16,

system to approach personalized driving behaviour, for emerging hybrid vehicles. The system is fully automated, non-intrusive with multi-modality sensing, based on smartphones. The application runs while driving and it will present personalized quantitative information of the driver's specific driving behaviour. In [18] a mobile application assesses driving behaviour, based on critical driving events, giving feedback to the driver. The Nericell system [15], from Microsoft Research, monitors road and traffic conditions using the driver's smartphone and corresponding incorporated sensors, but it can also detect honking levels and potholes on roads. The I-VAITS project [19] is an example that pretends to assist the driver appropriately and unobtrusively, analysing real-time data from the environment, from the car and from the driver itself, by the way the driver uses the different elements of the car, their movements or image processing of their face expressions. In [3], in the context of a car safety support system, an ambient agent-based model for a car driver behaviour assessment is presented. The system uses sensors to periodically obtain information about the driver's steering operation and the focus of the driver's gaze. In the case of abnormal steering operation and unfocused gaze, the system launches proceedings in order to slow down, stop the car and lock the ignition.

2 Related Work

Driving analysis can be a complex problem depending on the degree of information used and the number of categories being analysed. The analysis presented is described based on sustainable principles assessing driving patterns and their impact on sustainability and sustainable behaviour.

2.1 Driving Pattern Detection

Usually, driving pattern is defined and associated to the speed profile of the driver, but it can be expanded to other variables, as gear changing, and big changes on the acceleration [6]. In 1978, Kuhler and Karstens [12] introduced a set of ten driving pattern parameters. Later, in 1996, André [1] reviewed those parameters, and reviewed some of the most common parameters such as action duration, speed, acceleration, idle periods and number of stops per kilometre. In other studies [5], [6] other parameters were used to collect data from ordinary drivers in real traffic situations, such as wheel rotation, engine speed, ambient temperature, use of breaks and fuel-use. In these studies, GPS data was also monitored, where each driving pattern was attributed to street type, street function, street width, traffic flow and codes for location in the city (central, semi-central, peripheral). It was concluded that the street type had the most influence on the driving pattern. The analysis of the 62 primary calculated parameters, resulted in 16 independent driving pattern factors, each describing a certain dimension of the driving pattern. When investigating the effect of the independent driving pattern factors on exhaust emissions, and on fuel consumption, it was found that only 9 factors had a significant effect.

Table 1. Relevant attributes to driving analysis according to previous studies

Attribute	Ericsson [5]	Kuhler and Karstens [12]	Nericell [15]
Wheel rotation	+	+	-
Motor RPM	+	+	-
Pedals Monitoring	+	+	-
Street type	+	+	+
Fuel Consumption	+	+	+
Velocity	+	+	+
Acceleration	+	+	+
Standard deviation of acceleration	+	+	+
Trip duration	+	+	+
Hour of day	+	+	+

Table 1 provides an analysis of the main attributes identified. These studies share most of the identified attributes to analyse and classify driving patterns. With exception of attributes such as motor rotations per minute and pedals monitoring that are obtained directly from the physical vehicles, driving analysis with mobile and non-mobile sensors take interest in the same subgroup of attributes.

2.2 Sustainable Driving

Computational methods that allow the balancing of economic, environmental and social factors needed to a sustainable development, a newly emerging and interdisciplinary area, known as Computational Sustainability, solve problems which are essentially decision and optimization problems. The concept of sustainability and sustainable behaviours is important to ensure the welfare and well-being. Due to its importance, some researchers have discussed about quantification methods, and modelling sustainability [23], [11]. In the case of transportation systems, the assessment of the impact of a given driving pattern is made over sustainability factors, like fuel consumption, greenhouse gas emissions, dangerous behaviour or driving stress.

A system to estimate a driver profile using smartphone sensors, able to detect risky driving patterns, is proposed in [4]. It was verified whether the driver behaviour is safe or unsafe, using Bayesian classification. It is claimed that the system will lead to fuel efficient and better driving habits. In [9], and in addition to car sensory data, physiological data was continuously collected and analysed (heart rate, skin conductance, and respiration) to evaluate a driver's relative stress. The CarMa, Car Mobile Assistant, is a smartphone-based system that provides high-level abstractions for sensing and tuning car parameters, whereby developers can easily write smartphone applications. The personalized tuning can result in over 10% gains in fuel efficiency [7]. The MIROAD system, Mobile-Sensor-Platform for Intelligent Recognition Of Aggressive Driving [10], is a mobile system capable to detecting and recognizing driving events and driving patterns, intending to increase awareness and to promote safety driving, and thus possibly achieving a reduction in the social and economic costs of car crashes. The system uses Dynamic Time Warping and smartphone based sensor-fusion to detect and recognize actions without external processing.

3 Multimodal Assessment System

The implementation of the ubiquitous multimodal driving analysis system is depicted in this section. Ubiquitous monitoring is achieved by the use of smartphones equipped with accelerometer, GPS, compass, microphone and light sensors. These come as standard in most smartphones sold today. While it is not the main function of a smartphone, driving analysis can be achieved using some of the limited processing capability of low-end smartphones. Additionally, its connectivity options allow for better analysis on a server side location. The proof-of-concept system is illustrated by figure 1, where data flow is illustrated.

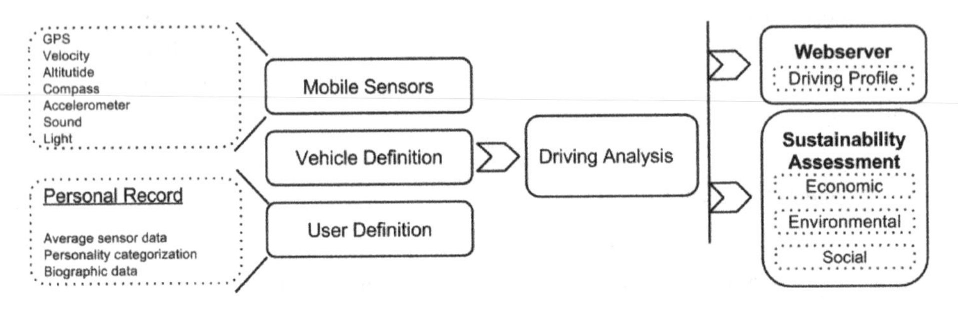

Fig. 1. Model describing the driving analysis in the system

After obtaining information about driving patterns, vehicle and driver definition, the driving analysis module will synchronize its data with an external server and derive a report assessment on the sustainable impact according to its 3 dimensions is built. Such knowledge is useful to update sustainability frameworks such as PHESS [21] which monitor and assess sustainable impact through performance indicators based on the same three categories.

The economic and environmental assessment is derived from the estimated impact of current driving patterns on the vehicle fuel consumption and gas emission. The social component is assessed by the effect of driving patterns on the social and psychological response from drivers. Although, sensor analysis by itself can answer if an event can or cannot be considered aggressive, it is still a reactive and instantaneous concept. With such information it is possible to map optimal to suboptimal configurations as well as infer which emotions these conditions produce on human beings. In a computational system it is still a challenging process to acquire rich information in this domain, but there are some approaches proposed in the research community that have had broad acceptance. In regard to the representation of a personality, the OCEAN structure used is similar to the approach adopted by the ALMA framework to represent the personality of people and initiate mood states in [8]. In the OCEAN approach, personality are defined by a set of variables (o, c, e, a, n) which represent five personality traits: Openness, Conscientiousness, Extraversion, Agreeableness and Neuroticism. On the other hand, the PAD space [14], is a computational friendly representation of mood states. A person's mood is represented using the three variables that define PAD space, Pleasure, Arousal and Dominance respectively. Contrary to the personality which is almost regarded as static during people lifetimes, mood is a temporal state of the human mind that can last for minutes, hours or even days.

3.1 User Driving Pattern

The analysis of driving patterns is made with the help of profiles. These structures are created individually for each driver and maintained in a web-server through the use of restful web-services. Although not using the internal data from vehicle sensor as in past research [5], the approach followed in this work uses smartphone data for ubiquitous and pervasive monitoring. An illustration of the application devised to record driver's attributes is detailed in the left part of figure 2. Data gathering is made through sensors, which is pre-processed internally with data fusion methodologies to enrich data and provide richer information. The number of variables used to assess driving patterns is based on the information gathered in the literature and adapted to ubiquitous sensors. As such, a total of 6 basic attributes are monitored: accelerometer, velocity, altitude, time of day, compass and position. From the fusion of these attributes, it is possible to infer standard deviations for each attribute according to each driver, the number of breaking and accelerations and its mean duration and intensity. These are the characteristics used in each driver profile in order to assess its regular driving behaviour. Asidc from the regular driving pattern, the system will also classify aggressive driving patterns, which are categorized by higher frequency of breaks and accelerations with high intensity and shorter duration than the driver regular behaviour. The rigth part of figure 2 shows a graphical

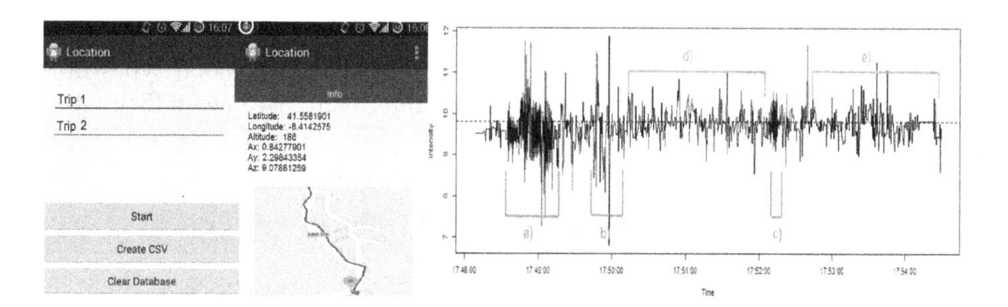

Fig. 2. Mobile Driving Patterns Extraction

representation of the intensity of the accelerometer sensor annotated with information gathered by other sensors according to driving pattern of a test driver. Event $a)$ represents low velocity and high variation of accelerometer which can be deduced to be parking or congested traffic driving. Events $b), c)$ represent aggressive events where velocity was kept high and sudden changes of direction result in high amplitude variation of accelerometer forces. Events $d), e)$ do not offer significant variation from common driving pattern thus identified as regular driving events. It is the identification of these events that will play a major role to detect the current mood of the driver. The pattern identified and its duration will provide information to assess the current driver's mood.

3.2 Driving Mood Classification

Mood classification is based on the analysis of the number of breaks and accelerations detected by mobile sensors: aggressive driving styles are connected with high frequency of breaking and accelerating actions; relaxed driving is correlated with stable velocity and low breaking and accelerating actions. The computational representation of mood states is done according to the Pleasure, Arousal, Dominance framework described by Mehabian (PAD) [14] and the PAD space extended by Gehbard in [8], where the initial P, A, D variables are initiated according to each user's personality assessment. The initial assessment of a driver's personality is achieved by an initial questionnaire, filled the first time the mobile sensing application is used. In this case the Newcastle Personality Assessor (NPA) questionnaire was used [16].

$$M_{actual} = M_{actual} + (M_{final} - M_{actual}) * emotionalWeight \tag{1}$$

Following this approach, two final states are defined representing aggressive and calm states so that a driver's mood may be updated towards one of these states. The equation 1 represents the current driver's state moving in a vector space at a velocity defined by the $emotionalWeigth$. This is dependent on values from the OCEAN personality representation and the assessment of the emotional response to a driving pattern. If positive update (relaxed pattern) $emotionalWeight = (e + 1.0) * const$ otherwise $emotionalWeight = (n + 1.0) * const$.

This study is limited to a classification of mood states oscillating between aggressive and relaxed moods. The classification of additional mood types increases not only complexity but also the error rate, as it becomes difficult to distinguish between them.

4 Conclusions and Future Work

In this article it is present an analysis of some state of the art related ubiquitous sensing of driving patterns. The approach with this work allows for pervasive monitoring through common objects such as smartphones in order to record and analyse driver's actions. Although limited by the sensors of each device, realist results can be extracted by standard hardware. The assessment of current driving mood is robust to sporadic deviant events as only a series of continuous negative or positive patterns does change the mood assessment while still reacting instantly to aggressive or relaxed events.

As a future work, the ubiquitous application will be extended allowing the sharing of information between drivers. With these abilities, gamification elements will try to moderate driving patterns by adding positive and negative points to a driver's profile. Additionally, the information shared allows for instantaneous information about nearby drivers such as emotion classification. The analysis will also add route classification based on an aggregated knowledge of driving patterns for each route, as well a It will also be used in future recommender systems and navigation applications. It will also allow automatic identification of driving actions through sensor analysis, thus requiring no user input at the start of driving records, enhancing the ubiquitous and pervasive nature of this work.

Acknowledgements. This work is part-funded by ERDF - European Regional Development Fund through the COMPETE Programme (operational programme for competitiveness) and by National Funds through the FCT – Fundação para a Ciência e a Tecnologia (Portuguese Foundation for Science and Technology) within project FCOMP-01-0124-FEDER-028980 (PTDC/EEI-SII/1386/2012). It is also supported by a doctoral grant, SFRH/BD/78713/2011, issued by FCT in Portugal.

References

1. André, M.: Driving Cycles Development: Characterization of the Methods. Tech. rep., INRETS (May 1996)
2. Aztiria, A., Izaguirre, A., Augusto, J.C.: Learning patterns in ambient intelligence environments: a survey. Artif. Intell. Rev. 34(1), 35–51 (2010)
3. Bosse, T., Hoogendoorn, M., Klein, M.C.A., Treur, J.: A Component-Based Ambient Agent Model for Assessment of Driving Behaviour. In: Sandnes, F.E., Zhang, Y., Rong, C., Yang, L.T., Ma, J. (eds.) UIC 2008. LNCS, vol. 5061, pp. 229–243. Springer, Heidelberg (2008)
4. Eren, H., Makinist, S., Akin, E., Yilmaz, A.: Estimating driving behavior by a smartphone. In: 2012 IEEE Intelligent Vehicles Symposium, vol. (254), pp. 234–239. IEEE (June 2012)
5. Ericsson, E.: Variability in exhaust emission and fuel consumption in urban driving. In: Urban Transport Systems, Proceedings from the 2nd kfb Research Conference, pp. 1–16 (1980)
6. Ericsson, E.: Independent driving pattern factors and their influence on fuel-use and exhaust emission factors. Transportation Research Part D: Transport and Environment 6(5), 325–345 (2001)
7. Flach, T., Mishra, N., Pedrosa, L., Riesz, C., Govindan, R.: CarMA. In: Proceedings of the 9th ACM Conference on Embedded Networked Sensor Systems, SenSys 2011, p. 135. ACM Press, New York (2011)

8. Gebhard, P.: ALMA: a layered model of affect. In: Proceedings of the Fourth International Joint Conference on Autonomous Agents and Multiagent Systems, pp. 29–36 (2005)
9. Healey, J., Picard, R.: Detecting Stress During Real-World Driving Tasks Using Physiological Sensors. IEEE Transactions on Intelligent Transportation Systems 6(2), 156–166 (2005)
10. Johnson, D.A., Trivedi, M.M.: Driving style recognition using a smartphone as a sensor platform. In: 2011 14th International IEEE Conference on Intelligent Transportation Systems (ITSC), pp. 1609–1615. IEEE (October 2011)
11. Kharrazi, A., Kraines, S., Hoang, L., Yarime, M.: Advancing quantification methods of sustainability: A critical examination emergy, exergy, ecological footprint, and ecological information-based approaches. Ecological Indicators, Part A 37, 81–89 (2014)
12. Kuhler, M., Karstens, D.: Improved Driving Cycle for Testing Automotive Exhaust Emissions. Tech. rep., Volkswagenwerk AG (February 1978)
13. Li, K., Lu, M., Lu, F., Lv, Q., Shang, L., Maksimovic, D.: Personalized Driving Behavior Monitoring and Analysis for Emerging Hybrid Vehicles. In: Kay, J., Lukowicz, P., Tokuda, H., Olivier, P., Krüger, A. (eds.) Pervasive 2012. LNCS, vol. 7319, pp. 1–19. Springer, Heidelberg (2012)
14. Mehrabian, A.: Pleasure-arousal-dominance: A general framework for describing and measuring individual differences in temperament. Current Psychology 14(4), 261–292 (1996)
15. Mohan, P., Padmanabhan, V.N., Ramjee, R.: Nericell. In: Proceedings of the 6th ACM Conference on Embedded Network Sensor Systems, SenSys 2008, p. 323. ACM Press, New York (2008)
16. Nettle, D.: Personality: What makes you the way you are. OUP Oxford (2007)
17. Oliveira, T., Novais, P., Jose, N.: Guideline Formalization and Knowledge Representation for Clinical Decision Support. Advances in Distributed Computing and Artificial Intelligence Journal (ADCAIJ) I(2), 1–12 (2012)
18. Paefgen, J., Kehr, F., Zhai, Y., Michahelles, F.: Driving Behavior Analysis with Smartphones: Insights from a Controlled Field Study. In: Proceedings of the 11th International Conference on Mobile and Ubiquitous Multimedia, pp. 36:1–36:8. ACM, USA (2012)
19. Rakotonirainy, A., Tay, R.: In-vehicle ambient intelligent transport systems (I-VAITS): towards an integrated research. In: Proceedings of the 7th International IEEE Conference on Intelligent Transportation Systems (IEEE Cat. No.04TH8749), pp. 648–651. IEEE (2004)
20. Sadri, F.: Ambient intelligence. ACM Computing Surveys 43(4), 1–66 (2011)
21. Silva, F., Analide, C., Rosa, L., Felgueiras, G., Pimenta, C.: Ambient Sensorization for the Furtherance of Sustainability. In: van Berlo, A., Hallenborg, K., Rodríguez, J.M.C., Tapia, D.I., Novais, P. (eds.) Ambient Intelligence & Software & Applications. AISC, vol. 219, pp. 179–186. Springer, Heidelberg (2013)
22. Sun, J., Wu, Z.H., Pan, G.: Context-aware smart car: from model to prototype. Journal of Zhejiang University Science A 10(7), 1049–1059 (2009)
23. Todorov, V., Marinova, D.: Modelling sustainability. Mathematics and Computers in Simulation 81(7), 1397–1408 (2011)

CloudFit: A Cloud-Based Mobile Wellness Platform Supported by Wearable Computing

Angel Ruiz-Zafra[1], Manuel Noguera[1],
Kawtar Benghazi[1], and José María Heredia Jiménez[2]

[1] Dpt. de Lenguajes y Sistemas Informáticos, University of Granada,
E.T.S.I.I., c/Saucedo Aranda s/n, 18071 Granada, Spain
[2] Dpt. de Educación Física y Deportiva, University of Granada,
Campus Universitario de Ceuta, c/Cortadura del Valle s/n, 51001 Ceuta, Spain
{angelr,mnoguera,benghazi,herediaj}@ugr.es

Abstract. Health and wellness area is an emerging social concern. The emergence of Cloud Computing and the growth of new technologies as smartphones and all kinds of wearable devices have given rise to delocalized health and wellness management systems and applications. Most of these systems, which are used by users on their own, are designed to track the exercises, monitor the physiological variables or as dietary diary with the aim to change the diary habits of users to improve their health and wellness, that could also be enhanced by the participation of expert advisors in the supervision of these activities.

This paper presents CloudFit, a mobile wellness platform supported by cloud technology and wearable devices, for supporting the monitoring of diary habits and improve the interaction between users and expert advisors. The development approach and design decisions taken for building CloudFit components are carried out by considering important characteristics of this kind of systems such as usability, accurate data capture and friendly data dissemination.

Keywords: cloud computing, soa, mobile computing, software components, ehealth, mhealth, wellness, wearable computing.

1 Introduction

In the last few years there has been a change of mind about the importance of health and wellness. Several organizations have focused their attention about health promotion, emphasizing on changeable health risk factors such as smoking, unhealthy eating or physical inactivity [1]. This change about the public attention to health promotion has also economic reasons, inasmuch as if the healthy lifestyle is not encouraged the costs of the unhealthy habits will be highly increased in the future [2-3].

This interest in healthy behaviors has permeated in society at all levels. From the nutrition education and healthy habits at schools to the continuous monitoring of blood pressure in elderly people or the importance of healthy lifestyle for athletes [4].

Physical activity has been shown to be an important factor related to a number of health outcomes [5]. The ability to measure physical activity behavior is useful, not

C. Ramos et al. (eds.), *Ambient Intelligence - Software and Applications*,
Advances in Intelligent Systems and Computing 291,
DOI: 10.1007/978-3-319-07596-9_17, © Springer International Publishing Switzerland 2014

only to understand the association between physical activity and health, but also for many other reasons, such as to monitor secular trends in behavior and to evaluate the effectiveness of interventions and programs [6]

The growth of new technologies such as smartphones and wearable devices (medical sensors, smart watch, smart glasses, body-sensors) [22], alongside with the emerging computer paradigms as Cloud Computing, have contributed greatly to lead a healthy life [7], thanks to the eHealth/mHealth application and systems supported by these technologies. Smartphones-internal sensors and the use of external wearable devices as data collectors are a powerful source of physiological, inertial and contextual information. The use of these sensors (GPS, vision sensors, audio sensors, light sensors, temperature sensor, acceleration sensor) open new opportunities for coach and sportsmen, increase the control of athletes and the following of physical condition in real time. Some of the advantage in the use of these sensors in the physical activity are: real-time tracking, monitoring physical variables, quantitative targets data, integration of different sensors and light device to use in sports environment, among others.

However, most of these systems are designed to be used by users on their own without help or expert advice, therefore usability is a major concern to address when constructing this kind of systems. This model usage is sufficient for most users, but for some kinds of users such as elderly people, people with eating disorders or athletes is needed an expert supervision to improve their health and wellness habits.

The supervision of expert advisors (trainer, dietitian, psychology, doctor, cardiologist) in the wellness habits of people can be enhanced by the use of the cloud, thanks to the easy and delocalized access to the information.

This paper describes CloudFit, a cloud-supported platform to monitor health and wellness of people. The platform is based in the use of smartphones and wearable devices and is designed to be used by different kinds of users, such as trainers, sport researchers, dietitian, elderly people and athletes, as well as in different contexts as telerehabilitation, physiological variables monitoring, evolution of athletes and track of exercises, among others. Some key design decisions have been taken due to the dynamic nature of wellness area, such as the enhancing of SOA or the use of a component-based platform [8] to ensure the adaptability over time. Moreover, features as usability and the good data management and processing are crucial topics addressed in the design process.

This paper is organized as follows. Section 2 presents related work. The platform, its features, architecture and applications are introduced in Section 3. Finally, Section 4 summarizes the conclusions and future work.

2 Related Work

Several systems have been proposed to improve health and wellness. These systems belong to a research projects as well as industrial ones.

Most of these systems are used as data collectors of the sensors of the mobile phone, as the use of GPS to know the distance traveled [9-10], the time spent in the exercise or the accelerometer values [11]. Other systems, instead, use external wearable devices such as GPS, heart rate sensors and accelerometers [12-14].

Other projects based on the use of smartphones and related with food control/eating habits [15] or monitoring caloric balance per day are gaining popularity recently [16].

Cloud technology is a useful tool to be used in this context, where several cloud-supported projects have been proposed. Data loggers of wearable devices [17] and monitoring systems [18] are the most common projects that use cloud technology.

In a commercial scope, many applications from different operating systems as Android or iOS are very popular nowadays. Applications such as [19-21] have millions of download and are used by all kind of people.

The users use these applications by their own to track the performance of the exercise and monitor the time spent, heart rate variability, know the route, food control, etc. Thus, the users themselves review their progress and adapt their habits.

This behavior entails that information such as physiological features or particularities of each user is not taken into account. The supervision of an expert advisor could overcome these issues.

The work presented in this paper intend to solve this lack of communication and interaction between experts and users, through a platform based on cloud technology to enhance this relations and mobile applications with wearable devices as data collector.

3 CloudFit: Cloud-Based Mobile Wellness Platform

In the following sections it is introduced the CloudFit platform, the features and principles that have guided its design.

3.1 Platform Overview

CloudFit is a cloud-supported platform to improve health and wellness of the people. The platform is made up by a set of cloud services, a mobile and web application and supported by wearable devices (Figure. 1).

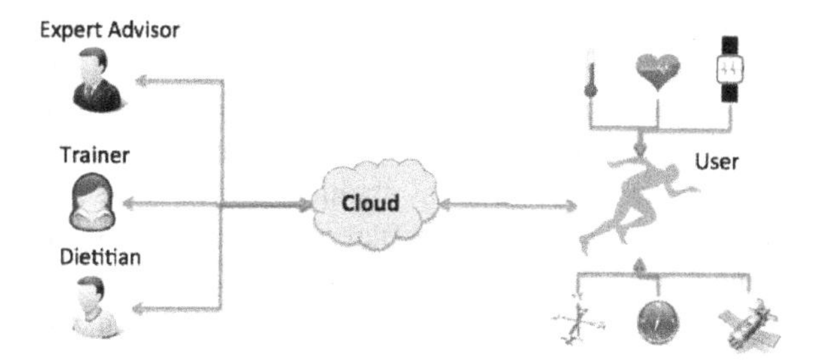

Fig. 1. CloudFit Scenario

The two main purposes of the platform are to:

- Monitor through the mobile application the values from wearable devices and several daily habits such as eating habits and exercises performing, and store and process this data correctly.
- Provide a tool for experts to supervise the habits of users and improve the communications between users and experts, with the aim of the improving their wellness, taking decisions based on the information from wearable devices, eating habits, the time spent on the exercises and user's feedback, among others.

The users use a mobile application to perform exercises (running, football, tennis) or daily activities (walk, eat), while they are using wearable devices from different nature as data collectors. The expert advisor or supervisor can monitor at real-time the performing of exercises through a web platform, manage the different exercises that users have to perform or communicate with user through a messaging system.

3.2 Platform Features

The platform must comply with certain features to reach to cover the purposes described above. The most important identified features are:

- Cross-platform messaging system, enabling the communication between expert advisors and users through different platforms (Desktop, Android, iOS).
- Timely alert; when a physiological value of certain user increase over medical boundaries, the system should react to notify it to the user as well as their supervisor.
- Multi-user views are supported by all the applications to allow the same application to be used by different kinds of users.
- Real-time monitoring by supervisors about the activities performed by users
- Seamless clouds, i.e., users (either expert advisors or users), are not aware that they are interacting with a cloud infrastructure
- Security and privacy. The system supports a one-way hash method as authentication method to ensure the authorized access to the API, as well as a bidirectional encryption method for the sensitive data in the database (email, name and lastname, messages).
- Interoperable system, in order to overcome the heterogeneity of system devices and enable the system to communicate and integrate with other systems. XML and JSON based representations of data in conjunction with some extensible and adaptable communication protocols are used to achieve this goal.
- Use of wearable devices. The system should be able to use different devices based on different technologies with different purposes. This feature guarantees the use of commercial and close-source hardware devices of popular vendors.
- The efficiency and robustness in the capturing data process from the sensors is a crucial aspect in this project to provide a reliable source of data. The use of dedicated algorithms to the large and consistent amount of data from the sensors stored in the cloud allows obtaining relevant and useful results.

3.3 Architecture

The health and wellness area is constantly evolving. New medical, health or sport requirements as well as novel areas may be included within this context. Furthermore, new wearable devices from several purposes and based in novel technologies will appear in the future, improving those that exists.

Thus, wellness systems should be adaptable over time to evolve in the same way, enabling that novel areas can be supported by these systems, expert advisors of these areas can use the system and new wearable devices can be integrated

The design decisions taken for the project presented in this paper have been made to ensure the easy evolution of the system. The solution focuses in the use of two different technologies: SOA (Service Oriented Architecture) and CBD (Component-Based Development).

The services oriented architecture designed and developed in this project is supported by cloud technology, to ensure primarily the scalability and the storage problems over time. A set of services have been developed to meet the different requirements of the system: management of physiological values, management of user information, algorithms to process values, management route of outdoor exercise, management of training, etc. Accordingly, new requirements or functionalities for specific experts advisors could be addressed adding new services to the architecture, ensuring the extensibility of the system. Also, the use of cloud-services ensures interoperability, enabling that different applications or systems from different platforms/OS can use these services.

The component-based development is represented in this project as a platform used by the mobile application [8]. This programming paradigm ensures the reusability of software components to construct new applications. Although currently there is only one application, the creation time of a new one is significantly reduced using the platform.

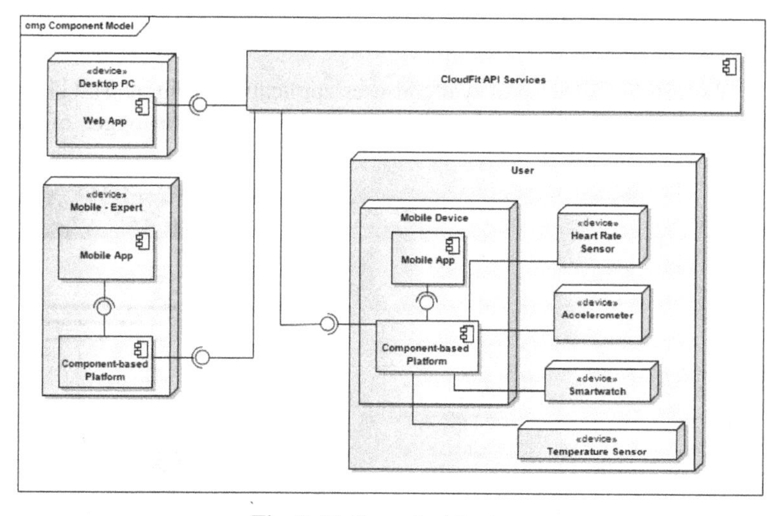

Fig. 2. Platform Architecture

In the same way, new components can be added to the platform, such as new components to manage new wearable devices or replace components to ensure the proper functioning. This solves the support for the future use of new technology in the system.

The architecture of the platform (Figure 2) is made up by several components and applications. The platform presented in [8] is used to provide different functionalities used by the mobile applications, and is responsible to manage the most of the functionalities presented in the mobile application: communication with the cloud, management of the sensors, data storing, receive of messages/notifications, etc. Furthermore, a set of cloud services gives support to the platform to provide several functionalities: recover information, user's management, etc. These services can be used by any system, application or platform, (mobile, desktop or web), promoting the interoperability.

3.4 Usability

In the creation process of this project, several meetings were held between computer scientists and sport science researchers to design the interface of the system applications, to ensure the usability of the system.

Furthermore of basic usability features such as consistent user interface, friendly ease of use or learnability, other usability issues to solve were identified: the touch gestures problems detected while the users are performing the exercises and the unable to view the smartphone screen because most of times is located in the arm are the mainly one's.

As solution, a custom gesture code proposed by sport science researches was supported by the application: one touch to receive the current state of physiological, inertial and contextual values or long touch to start/stop the monitoring are some of them.

3.5 Example of Applications

So far, the platform is made up of two end-user applications (Figure.3): (1) a mobile application used by users to perform the exercises and as data logger of different

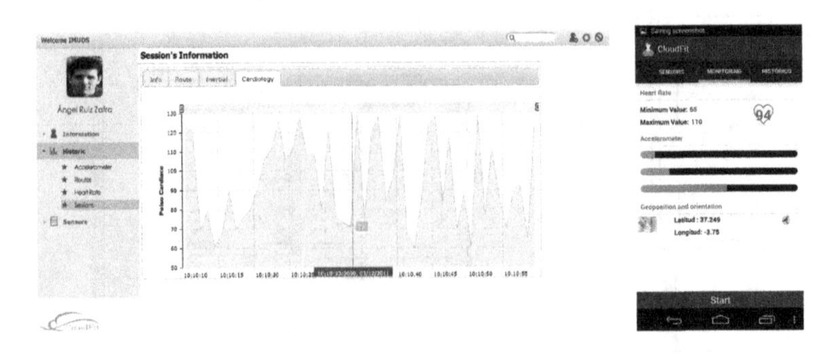

Fig. 3. CloudFit Web Platform (Left) and CloudFit Mobile Application (Right)

sensors; (2) a web platform used by supervisors to manage exercises or workout of users, monitor the physiological values, interact with uses through messaging system, etc.

The mobile application has been developed using the platform presented in [6]. Several components of the platform have been used. Each of these components is responsible for a specific task such as Bluetooth management. , internal sensor or cloud communications management.

4 Conclusions and Future Work

Nowadays, the people are more concerned about healthy lifestyle and wellness habits due to their proven benefits. Regular physical activity, diet or personal trainer are topics of interest.

The emerging of Cloud Computing and the growth of new technologies as smartphones and wearable devices (able to work as body-sensors) have boosted systems and applications dedicated to monitor their habits in different contexts (exercise, eating habits, physiology), improving their wellness. Furthermore, this improvement of wellness using these systems can be enhanced if expert advisors act as supervisors to change the habits in the proper way.

In this paper, a cloud-based mobile wellness platform has been described. The platform, called CloudFit, use smartphones and wearable devices as data collectors and has been designed taking into account the dynamic nature of these system, whereby two different approaches has been used: SOA and CBD. The aim of the platform is to monitor the daily activities (exercise, train, eating habits), the physiological, contextual and inertial values to improve the wellness. Furthermore, expert advisors are able to interact with the users to improve wellness and change their habits.

In order to achieve this, a services-oriented architecture supported by cloud technology has been presented, promoting the interoperability, extensibility and scalability of the platform.

As for future work, we are currently working in the improvement of the web platform to enhance the interaction between experts and users, the improvement of the mobile application adding new functionalities and the development for other platforms such as iOS. Also, the improvement of the integration, synchronization and management of wearable devices is another research topic to address.

Acknowledgments. This research work is part of the project "Sistema Ergonómico Integral para la evaluación de la locomoción como predictor de la calidad de vida relacionada con la salud en Mayores (Ergoloc)", funded by the Spanish Ministry of Economy and Competitiveness under the project DEP2012-40069 and by the Granada Excellence Network of Innovation Laboratories (GENIL) under project PYR-2014-5. The authors would also like to acknowledge contribution from COST Action IC1303.

References

[1] World Health Organization. Milestones in health promotion: Statements from global conferences (2009)
[2] Cecchini, M., Sassi, F., Lauer, J.A., Lee, Y.Y., Guajardo-Barron, V., Chisholm, D.: Tackling of unhealthy diets, physical inactivity, and obesity: health effects and cost-effectiveness. The Lancet 376(9754), 1775–1784 (2010)
[3] Colditz, G.A.: Economic costs of obesity and inactivity. Medicine and Science in Sports and Exercise 31(suppl. 11), S663–S667 (1999)
[4] Johansson, L., Thelle, D.S., Solvoll, K., Bjørneboe, G.E.A., Drevon, C.A.: Healthy dietary habits in relation to social determinants and lifestyle factors. British Journal of Nutrition 81, 211–220 (1999)
[5] U.S. Department of Health and Human Services. Physical Activity and Health: A Report of the Surgeon General. Atlanta, GA: U.S. Department of Health and Human Services, Centers for Disease Control and Prevention, National Center for Chronic Disease Prevention and Health Promotion (1996)
[6] Ward, D.S., Evenson, K.R., Vaughn, A., Rodgers, A.B., Troiano, R.P.: Accelerometer use in physical activity: best practices and research recommendations. Medicine and Science in Sports and Exercise 37(11), S582 (2005)
[7] Alex Mu-Hsing, K.U.O.: Opportunities and challenges of cloud computing to improve health care services. Journal of Medical Internet Research 13(3) (2011)
[8] Ruiz-Zafra, Á., Benghazi, K., Noguera, M., Garrido, J.L.: Zappa: An Open Mobile Platform to Build Cloud-Based m-Health Systems. In: van Berlo, A., Hallenborg, K., Rodríguez, J.M.C., Tapia, D.I., Novais, P. (eds.) Ambient Intelligence & Software & Applications. AISC, vol. 219, pp. 87–94. Springer, Heidelberg (2013)
[9] Eskofier, B., Hartmann, E., Kühner, P., Griffin, J., Schlarb, H., Schmitt, M., Hornegger, J.: Real time surveying and monitoring of athletes using mobile phones and GPS. International Journal of Computer Science in Sport 7(1), 18–27 (2008)
[10] Sandru, S., Hornos, M.J., Rodríguez, M.L.: Long-distance runner training system for smartphones. In: Proceedings of the 13th International Conference on Interacción Persona-Ordenador, p. 35. ACM (2012)
[11] McCarthy, M.W., James, D.A., Rowlands, D.D.: Smartphones: Feasibility for real-time sports monitoring. Procedia Engineering 60, 409–414 (2013)
[12] Kugler, P., Schuldhaus, D., Jensen, U., Eskofier, B.: Mobile Recording System for Sport Applications. In: Proceedings of the 8th International Symposium on Computer Science in Sport (IACSS 2011), Liverpool, pp. 67–70 (2011)
[13] Kwapisz, J.R., Weiss, G.M., Moore, S.A.: Activity recognition using cell phone accelerometers. ACM SIGKDD Explorations Newsletter 12(2), 74–82 (2011)
[14] Depari, A., Flammini, A., Rinaldi, S., Vezzoli, A.: Multi-sensor system with Bluetooth connectivity for non-invasive measurements of human body physical parameters. Sensors and Actuators A: Physical (2013)
[15] Six, B.L., Schap, T.E., Zhu, F.M., Mariappan, A., Bosch, M., Delp, E.J., Boushey, C.J.: Evidence-based development of a mobile telephone food record. Journal of the American Dietetic Association 110(1), 74–79 (2010)
[16] Lee, G., Tsai, C., Griswold, W.G., Raab, F., Patrick, K.: PmEB: a mobile phone application for monitoring caloric balance. In: CHI 2006 Extended Abstracts on Human Factors in Computing Systems, pp. 1013–1018. ACM (2006)

[17] Rowlands, D.D., McNab, T., Laakso, L., James, D.A.: Cloud based activity monitoring system for health and sport. In: The 2012 International Joint Conference on Neural Networks (IJCNN), pp. 1–5. IEEE (2012)
[18] Serhani, M.A., Benharref, A., Badidi, E.: Towards dynamic non-obtrusive health monitoring based on SOA and cloud. In: Huang, G., Liu, X., He, J., Klawonn, F., Yao, G. (eds.) HIS 2013. LNCS, vol. 7798, pp. 125–136. Springer, Heidelberg (2013)
[19] `https://www.runtastic.com/`
[20] `http://www.endomondo.com/`
[21] `http://www.myfitnesspal.com/`
[22] Mann, S.: Smart clothing: The shift to wearable computing. Communications of the ACM 39(8), 23–24 (1996)

Developing Ambient Support Technology for Risk Management in the Mining Industry

Helena Lindgren[1,*], Lage Burström[2], and Bengt Järvholm[2]

[1] Department of Computing Science,
[2] Occupational and Environmental Medicine,
Department of Public Health and Clinical Medicine
Umeå University, SE-901 87 Umeå, Sweden
{helena.lindgren,lage.burstrom,bengt.jarvholm}@umu.se

Abstract. There is a major goal in the mining industry to reduce risks and maintain health in work environments. Moreover, the industry is obliged to monitor the risks in work environment as well as employers' health statuses. The potentials in using ambient information for the purpose to reduce risks, prevent work-related injuries and monitor health in individuals has been explored. Applications tailored to the individual are being developed to aid the worker in mining or mining-related work environments in valuing the risks of their work situation and create awareness in the individual about how he or she can decrease risks for primarily physical damages. The purpose is to encourage the worker to act upon the level of risk for injuries, and upon the new insights the worker gain from the applications. The identified opportunities for and obstacles to integrating ambient information in these health applications are discussed.

Keywords: Ambient intelligence, Occupational health, Mining industry, End-user development, Behavior change systems.

1 Introduction

In this paper we present a study of a design and development process where medical domain professionals collaboratively design and implement Semantic web-based applications aimed at changing mining and construction workers awareness about risks in their work environments and encouraging the worker to act in order to decrease e.g., their exposure to risks. The goal of the applications is to empower the worker to take control over their work situation and becoming able to improve their work situation. A secondary goal of one of the applications is to allow in-house medical service providers to tailor computer-based health check ups to local work environments and individuals.

In addition to the hands-on development and modeling of available evidence-based medical knowledge, national regulations and adaptable user models, the possibilities to integrate ambient information obtained in the work environment

* Corresponding author.

C. Ramos et al. (eds.), *Ambient Intelligence - Software and Applications*,
Advances in Intelligent Systems and Computing 291,
DOI: 10.1007/978-3-319-07596-9_18,

was explored. This was done for enhancing and situating the tailored support to an individual so that the worker's current work environment is included in the assessments of risk exposure, and in the generation of tailored advice about how decreasing the exposure.

In an initial phase the exposure to vibrating machines and vehicles, dust and particles, which may give skin-related problems were in focus. Since the mining industry in the region in focus has routines for measuring different environmental factors regularly, we focus in this paper on the mining industry. We discuss the results from the different perspectives of the stakeholders. The potentials and obstacles will be discussed.

2 Methods

A participatory action research methodology is applied in the project (e.g., [1, 2]). The presented results were obtained during a period of iterative development, which included three phases of evaluation studies with potential end users. The periods between these phases were dedicated design and knowledge modeling work, involving medical and health professionals (both researchers and clinicians), and researchers in knowledge engineering and interaction design.

One particular focus in the design process was on motivational factors in individuals and how these could be mirrored, triggered and emphasized in the applications. Motivation was seen as a major drive to conduct risk assessments and risk management. The domain professionals utilize their own experiences from daily practice in motivating their patients or clients to change behavior. They also utilize research in environmental medicine and occupational health topics such as changing work routines and applying methods to monitor work related health issues (e.g., [3, 4]). They brainstormed ideas, limited the content to the most important and the most likely to have effect on behavior, and they sketched different ways of visualizing feedback to the user. This was done in numerous iterations with evaluations done within the group. Some of the ideas were implemented in prototypes by the physicians using the knowledge modeling application ACKTUS [5], and evaluated with potential end users. Use sessions were video recorded, participants were observed and interviewed, and notes were taken.

An initial evaluation study of the Vibration application was done with a group of eight domain experts. The results from the study were fed into further development of the application. Particular focus was the design of the questions, their answers, how the advices are perceived and the general design of the application.

Another evaluation study consisted of meetings with representatives from the in-house health care in each of the two participating mining industries. The purpose was to evaluate the initial Health Checkup prototype and put the purpose of the applications into the context of the respective organizations activities related to risk assessment and management. Design suggestions were generated and discussed.

A third evaluation study was conducted during two days at one of the in-house health care organizations where the prototype applications were integrated as part of ordinary health care work. The dialogue-based version of the Health Checkup application was used by the workers as part of the normal health screening routine as a replacement of the paper-based forms. The workers used a touchpad to interact with the web-application and filled in the requested information. During these days the design process continued with an outlining by the health professionals of potential extensions to the Health Checkup application for the purpose to achieve tailoring of the checkup and the advice related to risks provided the worker.

The results of the evaluation studies were analyzed from the perspectives of a theory of motivation, the Self-Determination Theory (SDT) [12], and a framework for persuasive technology [7], which were considered relevant for the targeted problem domain and work environment, further described in Section 2.2.

2.1 Material

The prototype applications (Dust Demo, Vibration and Health Checkup) were developed using ACKTUS [5] (Figure 1). ACKTUS is an evolving semantic web application that is designed to allow domain experts who are typically not familiar with knowledge engineering to author and model the knowledge content of, and design the interaction with, knowledge-based applications [5]. The results can be tested in a prototype end user application called ArbetsVis so that the domain experts can immediately see how the results will appear to a user.

For assessing and computing risk levels, measurement data from a database of vibrating machines was imported and modeled into a dedicated machine ontology in order to reuse the information in interaction with the domain experts and with end users. Dust measures were also included and ontologies of work tasks and professions were created. This generic background data is aimed to be supplemented with ambient information obtained in a potentially hazardous work situation particular to an individual worker.

2.2 Theory for Persuasive Technology and Behavior Change Systems

The benefits of using persuasive technologies are demonstrated in applications aiming at changing behavior and attitudes, such as quit smoking, reduce unhealthy food intake, increase physical exercise, etc., [8–10]. In the case of risks in the mining industry, a large proportion of accidents can be prevented by maintaining safe routines, which are dependent on the individual worker's awareness and attitude to risks. Consequently, persuasive technology may fill a purpose. However, there are numerous studies where the obstacles to changing behavior are described [10, 11, 4]. A profound knowledge about social, psychological, behavioral and environmental factors are necessary in order to design, evaluate and successfully implement persuasive technology (e.g., [12]). This knowledge is typically found among medical and health care personnel and social workers that meet clients in daily work where a significant part of their work aims at changing behavior to increase health. Consequently, they have a well-founded knowledge about the potential users of applications and their work and life situation.

The framework for persuasive technology presented in [7] identifies three factors that need to be present in order for an individual to change behavior and perform a target activity. The individual needs to be sufficiently motivated, have the skills to perform the activity/behavior and be triggered to perform the activity/behavior. Timing of the three is important. Three core motivators are described with a dialectical character: pleasure vs. pain; hope vs. fear; and social acceptance vs. rejection. In addition, facilitators are identified as part of the framework, which increase the ease by which an activity can be performed.

The Self-Determination Theory SDT [12] was used for analyzing the motives for the users to use applications modeled in the design process, and the data from evaluation studies with end-users. SDT distinguishes between intrinsic motivations (internal within an individual) and extrinsic motivations (evoked by sources external to the individual), which gives a framework for assessing potential reasons for activity. SDT identifies three needs as driving forces for activity: *relatedness*, *autonomy* and *competence*. In short, they cover the individual's need for being a part of a social context, having control over ones life situation and having the skills that are needed to be able to affect a situation.

2.3 Participants

Two domain professionals who are physicians and experts in two sub-fields of environmental medicine with more than 20 years of experience of treating patients from the mining and construction industries are leading the development of the different applications. They conducted the hands-on modeling of the content and the modeling of the interaction with the knowledge content using ACKTUS. They also participated in workshops with representatives from different industries, where the potentials for the integration of ambient information was investigated.

Eight medical domain experts who are treating patients with work-related medical problems in their daily work and had experiences with vibrating machines participated in an evaluation study of the Vibration application.

Additional two physicians and four nurses organized by two different mining industries participated in evaluation studies and contributed to the content of the Health Checkup application. The content consists of a set of common questionnaires that are composed and used by these organizations and some questions customized to each organization. One purpose is to make these tailored to individuals, so that relevant follow-up questions are presented, and so that the individual gains something when participating in the health checkup, e.g., in the form of tailored advice and support in how improving their work environments.

Five workers (four male and one female between the ages 35 to 60) employed by one of the mining companies participated in the pilot evaluation of the Health Checkup application and four workers (male between the ages 30 to 60) tested the Vibration application. Only one of the workers was experienced with touch pads, and they were all considered low or moderately skilled in computer use.

None of these participants were experienced with knowledge engineering or interaction design.

3 Results and Discussion

The results are discussed along the following themes: empowerment and motivational factors in end users, balancing ethical aspects with health and productivity, and human resources perspective and cost.

3.1 Empowerment and Motivational Factors in End Users

The domain experts approached the design task by taking as starting point how they typically conduct clinical interviews and assessments in dialogues with patients. The flow of interaction they implemented in the system was aimed to mimic how the physician typically conducts an investigation with the appropriate responses given as motivations to the patient to take their situation seriously and to do something about it. Based on this, they reduced the amount of necessary information to a minimum to optimize the effect - a potential change by the worker of the workers situation, or health status by visiting a physician for investigation.

The systems responses to the user was altered during the iterations, changing from general information to specific information addressed to particular users, such as advice to make contact with health care about increasing symptoms for a medical investigation, and/or motivations how the user can change work routines and environment in order to decrease the risk of developing more symptoms. Increased emphasis was put on content, information and advice tailored to individuals.

The strategy applied by the domain experts to attract interest and increase motivation was to provide the user a calculation of their risks to develop injuries presented in an easy to understand graphical way so that the user immediately understands the level of danger. Based on their current work situation suggestions of alterations are presented that improves the situation. The results are also visualized in a graphical form, showing how much the work environment could be improved based on the same algorithm that calculates the risk (Fig. 1).

Simplicity was strived for in the design process, and the timing of the three factors motivation, ability and trigger, was taken into consideration and accomplished in the health checkup situation. What was observed in this particular use situation was that the participants were unexpectedly positive to the use of the applications and expressed no stress about time or potential lack of ability in SDT-terms. A general positive attitude was observed, both towards testing the applications and towards their physical and social work environments. This was explained by the fact that the workers were obliged to do the health checkup as a part of their work, the time allocation was regulated outside of their control, thus the activity was extrinsically motivated. In addition, they acted as part of a community, in an activity organized by highly trusted individuals who were well known in their work environment. When the company according to national regulations routinely measure exposure to dust, workers are carrying measurement equipments during a time period, which is also considered as a part of their work tasks. However, currently the data is not used for calculating risks at an individual's level. It can be assumed that an individual worker may be more motivated to carry the measurement equipment if it was possible to also receive computed information in the perspective of the individual's profile, e.g., already accumulated exposures, which may lead to high levels shown in blood samples.

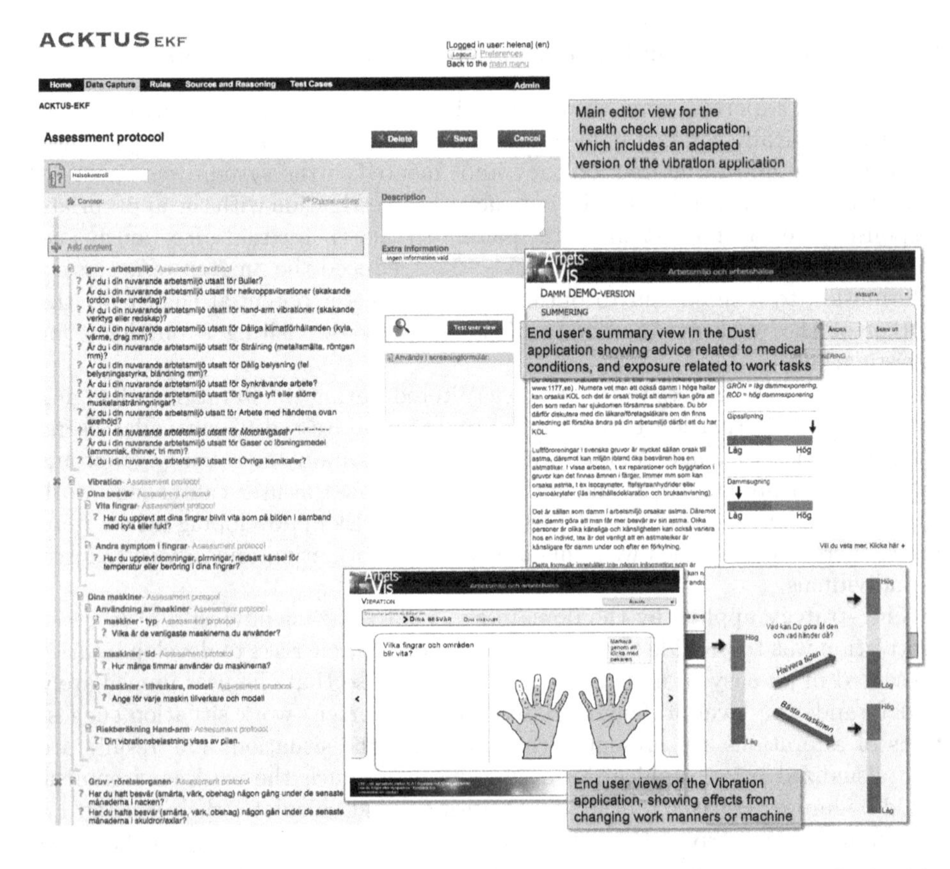

Fig. 1. Screenshots showing i) the ACKTUS editor where the content of the health check up application is composed by the physicians, ii) the summary view of the Dust application containing advices related to the user's medical condition and exposure in two different work tasks, iii) a question about white fingers, and iv) suggestions about how changing the work environment and the organization of work in order to reduce exposure to vibration.

3.2 Balancing Ethical Aspects with Health and Productivity

There were discussions from an ethical perspective in the design process about whether the potential evoking of fear of risks, was motivated by the potential gains with the applications. A worker may be motivated by the possibility to decrease pain, gain hope to improve their work environment so that they may be able (in terms of SDT [12]) to continue being a productive colleague in a collaborative work environment where work is heavily dependent on teamwork (social acceptance vs. potential rejection by their work community). Since the frustration typically was high in the individuals the physicians meet who are suffering from injuries and from not being able to work in the way they would like to be able, it was agreed that the gains were significant. This view was shared with the employers, who are dependent on their workers to be productive. Consequently, since the measurements done of the individual's physical status, the individual's physical work environments are regulated by national policies, and they are done for the purpose to protect the worker, the baseline attitudes are positive towards using the data for this purpose. However, another important factor is whether the worker actually trust their employer to take action to improve the situation, and/or allow the worker to adjust the work situation in case a dangerous situation is detected, e.g., through the use of an ambient system. It is likely that in some situations short term production goals are allowed to override the long time goals to maintain workers' health, even if health economic calculations show gains from a longer perspective.

3.3 Human Resources Perspective and Cost

In dialogues with the mining industry representatives a vision emerged of synthesizing a health checkup application with the specific applications such as the Vibration application and additional applications partly developed for dust and skin related issues. In the use of the health checkup application the medical professionals perceived the nesting of specific applications such as the vibration application to screening questions in the health checkup, and which leads to specific investigation of this as part of the health checkup, as being a way to rationalize their work and speed up their follow up activities on issues. In addition, the summary view that the nurses would like to see would highlight the deviating answers that the nurse would like to follow up in a dialogue with the worker.

Another line of development proposed by the participating professionals is to integrate environmental information other than the information about vibrating machines. The mining industries are measuring presence and quantity to e.g., dust in their work environments. In addition, data is collected about individuals' levels of particles in e.g., their blood, which is another indication of exposure. National regulations determine when a worker has too high levels and has to be taken out of a particular environment and work tasks. If the information is combined, the support can become tailored to individuals in a way that the individual can avoid particularly hazardous situations likely to increase risks above thresholds in this individual. This facilitates the development of a proactive ambient application that alerts the worker in cases of high exposure and may provide well-founded arguments, also from a health economic perspective, about what to do in a potentially hazardous situation.

4 Conclusions

Three applications have been designed and partly developed for risk assessment and risk management. The development process has been end-user driven and collaborative, involving medical domain experts and health representatives from the mining industry. The purpose of the applications is to increase mining and construction industry workers knowledge about risk factors in their work environments, their own complaints and how they can change their work environments. Ideally, the worker should also act upon the knowledge and change their work methods or environment. To motivate the worker to do this, domain experts have analyzed motivational factors and modeled motivations as tailored advice into dialogues with the user. A shift of focus was observed in the process, from general information-based content towards advice and motives tailored to the individual workers situation. Another shift was seen from the initial view that the worker was the main user of the applications, to also include health care personnel working on behalf of the company, as collaborators working towards improving health and work environments.

One conclusion that was made was that persuasive technology benefits from allowing domain professionals and domain experts design and model the content as well as the interaction with the content in the process of development. This increases the ecological validity and facilitates the management of the applications. The semantic web prototype application ACKTUS has been instrumental in the process and has also been developed as a result of the process.

Another conclusion was that the applications need to be integrated into the work environments, in both the routines of delivering the health care services, and in the production line. This for the purpose to optimize the tailored advice based on both domain knowledge, knowledge about the individual and knowledge about the physical and social environment, partly obtained through ambient systems. The ambient information currently obtained for filling the company's own purposes, can be refined and synthesized with individual information, for providing both the worker and the company motivation for a personalized risk management integrated in daily work. Moreover, the gains for the company in cost reduction for health care services, and for decreasing the need for replacing personnel can be optimised. Consequently, ongoing work includes the integration of environmental information measured in the mining industry work environments, which will be combined with the self-assessed information and the knowledge base in order to guide interventions.

Acknowledgments. The authors are grateful for the engagement from industry representatives throughout this project. The project was partly funded by AFA Försäkring and the Swedish Innovation Agency (Vinnova).

References

1. Davidson, E., Heslinga, D.: Bridging the IT Adoption Gap for Small Physician Practices: An Action Research Study on Electronic Health Records. Information Systems Management 24(1), 15–28 (2007)
2. Simonsen, J., Hertzum, M.: Iterative Participatory Design. In: Simonsen, J., Brenholdt, J.O., Büscher, M., Scheuer, J.D. (eds.) Design Research: Synergies from Interdisciplinary Perspectives, pp. 16–32. Routledge, London (2010)
3. Pettersson-Strömbäck, A.: Chemical exposure in the work place: mental models of workers and experts. Umeå, UmeåUniversity (2008)
4. Pettersson-Strömbäck, A., Liljelind, I., Neely, G., Järvholm, B.: Workers' interpretation of self-assessment of exposure. Ann. Occup. Hyg. 52(7), 663–671 (2008)
5. Lindgren, H., Winnberg, P.J., Winnberg, P.: Domain Experts Tailoring Interaction to Users – An Evaluation Study. In: Campos, P., Graham, N., Jorge, J., Nunes, N., Palanque, P., Winckler, M. (eds.) INTERACT 2011, Part III. LNCS, vol. 6948, pp. 644–661. Springer, Heidelberg (2011)
6. Fogg, B.J.: Persuasive Technology: Using Computers to Change What We Think and Do. Morgan Kaufmann Publishers, San Francisco (2003)
7. Fogg, B.J.: A Behavior Model for Persuasive Design. In: Persuasive 2009, April 26-29. Claremont, California (2009)
8. Consolvo, S., Everitt, K., Smith, I., Landay, J.A.: Design requirements for technologies that encourage physical activity. In: Proc. The Conference on Human Factors in Computing Systems (ACM SIGCHI), pp. 457–466 (2006)
9. Purpura, S., Schwanda, V., Williams, K., Stubler, W., Sengers, P.: Fit4life: the design of a persuasive technology promoting healthy behavior and ideal weight. In: Proc. of the Conference on Human Factors in Computing Systems (ACM SIGCHI), pp. 423–432 (2011)
10. Colineau, N., Paris, C.: Motivating reflection about health within the family: the use of goal setting and tailored feedback. UMUAI 21(4-5), 341–376 (2011)
11. Oinas-Kukkonen, H.: Behavior Change Support Systems: A Research Model and Agenda. In: Ploug, T., Hasle, P., Oinas-Kukkonen, H. (eds.) PERSUASIVE 2010. LNCS, vol. 6137, pp. 4–14. Springer, Heidelberg (2010)
12. Ryan, R.M., Deci, E.L.: Self-determination Theory and the Facilitation of Intrinsic Motivation, Social Development & Well-being. American Psychologist 55, 68–78 (2000)

Wireless Sensor Networks to Monitoring Elderly People in Rural Areas

Gabriel Villarrubia, Juan F. De Paz, Fernando de la Prieta, and Antonio J. Sánchez

Department of Computer Science and Automation, University of Salamanca
Plaza de la Merced, s/n, 37008, Salamanca, Spain
{gvg,fcofds,fer,anto}@usal.es

Abstract. Elderly residents who require personalized attention specific to their age-related needs generally inhabit rural areas. Users who are not of a very advanced age require only basic assistance, most commonly simple reminders to avoid forgetting or distractions. This article proposes the concept of a Smart City, which focuses on rural areas and incorporates a system to monitor and assist people of an advanced age who require a type of support which is usually complicated to provide in areas located far from urban centers. The proposed system incorporates WiFi networks, beacons, set topboxes and virtual agent organizations, and uses applications, interactive TV programs and a Wi-Fi based tracking system to monitor patients.

Keywords: Multiagent systems, wireless sensor network, health care.

1 Introduction

Rural areas are usually far from city centers and tend to be inhabited by elderly residents whose state of health may require basic monitoring and tracking [1] [17] [23]. There are certain services that provide direct communication to control centers through the use of alert buttons, thus providing communication for specific incidents[2] [3] [4] [11] [15]. Additionally, according to studies in this field [6], there have been many advances in areas such as telemedicine, the result of a continuous effort which can be observed in the evolution of certain devices [7]. Until now, almost all systems have focused primarily on a home environment, which requires bandwidth data connections that simply do not exist in many rural centers. This inexistence stimulated the need to create a system that could offer more extensive tracking of elderly residents, allowing them to easily interact with a television, and facilitate some of their daily tasks such as: making a doctor's appointment, selecting transportation time, confirming information from the local authorities, etc.

Current tracking systems used with elderly patients require the installation of sensors to monitor basic activities [2] [16] [18][38], and require communication networks to transmit information. In other cases, the requirements have more to do with monitoring aspects related to the environment[19], such as temperature [3], or those more specific to specific patient data such as oxygen level or pulse rate [4], even the

C. Ramos et al. (eds.), *Ambient Intelligence - Software and Applications*,
Advances in Intelligent Systems and Computing 291,
DOI: 10.1007/978-3-319-07596-9_19,

location of the patient [5] [11]. These systems tend to focus on specific medical aspects and require a very specific type of device [4], often influencing the normal behavior of the user. The technology used to analyze the information varies, although systems tend to focus on the use of data mining techniques [21][26][28][29] or artificial intelligence [31][35][40] such as neural networks [8][24][25][32][41] or multi-agent systems [9][20][30] [34][37][42].

Commercial systems require the installation of a large number of systems, which is costly and often a prohibitive factor in their use. This research group has already performed studies in assisting the elderly [11] [14]. The article proposes a system that integrates WiFi networks deployed in rural areas and facilitates communication of systems which are of low cost to the user since they do not require a private connection. A tracking and control system is created over the WiFi network by integrating a WiFi tag or mobile telephone to locate users in their homes and in specific nearby areas which they commonly frequent. An intelligent system with personalized multimedia content is used to control the user; this is done through a topbox which is connected to a television and offers continuous monitoring of the users in their homes to assist, for example, in reminding the patient to take pills, alerts to close an open window, or simply ask the user if they have a specific medical problem. The virtual organization of agents integrates intelligent algorithms to track and control the patient and send an alert when the patient engages in anomalous behavior.

This article is structured as follows: section 2 includes information about the proposal and the different techniques applied to monitor users; section 3 provides a case study, and section 4 presents the results and conclusions.

2 Proposed System

The proposed system is composed of a virtual organization [33][39] of agents which can carry out different tasks such as the localization and monitoring of users. The system was developed over the PANGEA architecture which offers possibilities such as the inclusion of light agents in different devices, for instance a topbox set. The architecture that the platform follows can be seen in Figure 1. As shown, there are two suborganizations, one associated with localization and another with processing the information obtained from the topbox set, which is in turn combined with localization.

2.1 Passive Localization

Passive localization is done by using the ddwrt operating system to modify the firmware in routers. Routers scan the users associated with access points as well as those not associated. Each router retrieves the information from the RSSI level measurements and sends them to a server in charge of locating the users according to the levels detected. The process is shown in Figure 2.

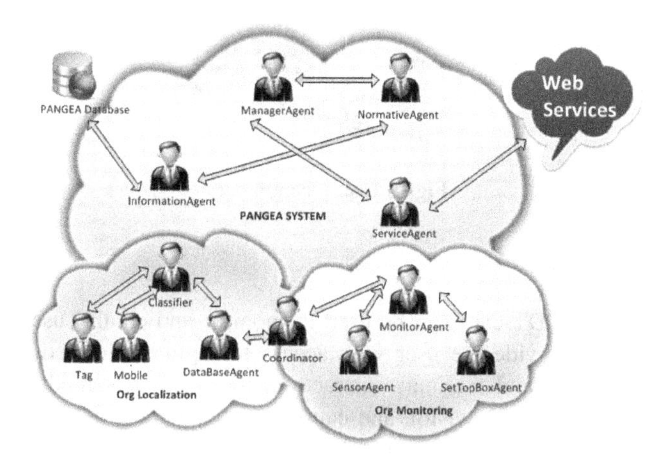

Fig. 1. Virtual organization of agents

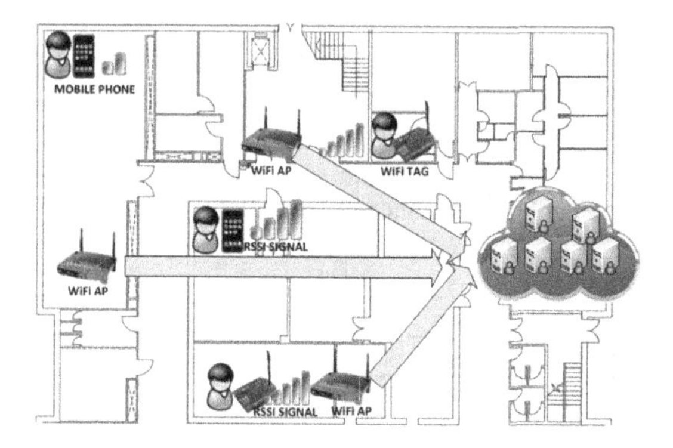

Fig. 2. Localization system architecture

In order to carry out this process, an automatic calibration is performed by a mobile phone. The mobile telephone equipped with a GPS is turned on, WiFi is activated, and position data is sent continuously to a server, which in turn cross references the position information with the RSSI level signals that were received.

Latitud, longitud, MAC BSSID antena1, RSSI antena2, … Latitude, longitude, MAC BSSID antena1, RSSI antena2, …

Based on this information, a training is performed by applying a SVM; the trained SVM is then used to determine the position of the user according to the signal levels detected by the WiFi antennas, without using the GPS. This makes it possible to use any WiFi device to locate users without needing to have a GPS, which in turn increases battery life. The passive localization was done by using the WiFi device seen in Figure 3. This allows for greater batter life and can be used in bracelets. Similarly, the user's mobile device can also be used.

Fig. 3. User location tag

2.2 Monitoring Users

A topbox set with a TDT card can be used to monitor and control users through the television and still provide the user with access to regular televised broadcasts. In order to input content and facilitate interaction, the VLC was modified to display messages while watching television and capturing the interaction of users in turning on and off the device. The messages and data retrieved are listed in Table 1.

2.3 Detecting Anomalies

Detecting anomalous behavior in a user can be done by observing the user's behavior as they interact with the system. User interactions are grouped into categories as indicated in Table 1.

Table 1. User interactions with the system

Turn on	Time
Turn off	Time
Pressing remote control	Time
Interactive response	Time
Question with alarm	Answered

The data gathered in Table 1 provides the information that will be used to determine normal user behavior. Three different procedures are available to determine anomalous behavior:

- Predefined rules: a rules system based on drools makes it possible to predefine rules according to the data registered in the data base. The rules are defined according to a set of conditions that when true will result in the execution of specific predefined actions.
- Interval-based: CBR [22][26][36] is used to determine a confidence interval for the values that have been detected so that the detection of a value outside the range will initiate an alert.
- Classifier: An SVM is executed within a CBR according to the cases considered anomalous and normal; each situation that is detected is classified as one of these two cases.

The predefined rules are shown in Figure 4. As we can see, they are simple and identify a condition and an action that is executed when a given condition is true. The rules engine probes the rules and activates them automatically when the condition is true. Additionally, it can modify the rules in execution time.

```
rule "Sleep problem"
    when
        $data : Data($data.getNumberOfTimes()>Configuration.getThresholdWalkNight
            && $data.getNode()==new User($data.getUserId()).getIdNodeBed())
    then
        ActionAlarm.sendMessageSleepProblem($data);
end
```

Fig. 4. Rules to determine unusual user behavior

The process for determining the confidence interval consists of grouping the days of the week by similarity and then creating confidence intervals. The definition of the cases are established as indicated in Table 2.

Table 2. Definition of a confidence interval

Turned on	Length of time turned on during interval
Number of times turned on	Number of times topbox set is turned on
Number of times turned off	Number of times topbox set is turned off
Remote control pulses	Number of remote control pulses during interval
Interactive pulses	Number of interactive pulses with remote control during interval
Alarm pulses	Number of pulses with alarm
Day of the week	Day of the week
Holiday	Yes or no
Time frame	Day Schedule interval

A two-way ANOVA with repetition is used to group the days of the week. The two factors are the day of the week and the time interval. The model that must be followed is shown in (1).

$$y_{ijk} = \mu + \alpha_i + \beta_j + (\alpha\beta)_{ij} + e_{ijk}$$

A day is divided into 24 intervals of one hour. For each group the days considered different, a confidence interval is calculated for every time interval. The complete process is described in the algorithm shown in Figure 5.

The set of rules based on confidence intervals is analyzed each time new data is received from either those shown in Table 1 or at the end of each time interval. In this case, the intervals are established as one hour.

Finally, in order to obtain a more advanced procedure to determine unusual user behavior, any anomalous behavior engaged in by the user is registered in the data base. This data base stores the information referring to the information shown in Table 3. Based on this information, an LMT [13] is trained to be able to detect anomalous behavior. When the probability of an anomaly is greater than a determined threshold, the user's state will be validated.

```
Input: V
Output: G, I
// V anlyzed variable
// G groups
// I confidence interval

G ← ∅;
I ← ∅;
k = 0;

g_k ← monday;
for dia_i ← monday to saturday do
    for dia_j ← dia_i + 1sunday do
        if anova_{g_k j v} accept H0: (αβ)_{ij} = 0 then
            g_k ← g_k ∪ dia_i;
        end
        else
            k ← k + 1;
            g_k ← dia_i;
        end
    end
end

G ←;
for i=0 to G do
    I ← I∪ confidence interval for the average α = 0.05;
end
```

Fig. 5. Calculation of confidence intervals

Table 3. User interactions with the system

Location	Zone (nearest antenna)
Signal level	Signal level from nearest antenna
Fields, Table 2	...
State of Alarm	Activate or Deactivate

3 Case Study

The system was tested in a small locality in the province of Salamanca (Spain). The WiFi networks were deployed using 120º sector panel antennas placed in blocks of 3 to cover the full 360º and provide full coverage; one of these antennas was placed in the center of the locality. For more distant areas, a sectorial antenna was used as it permits a client connection with the antenna base. The antenna used a repeater to create a virtual WiFi and extend coverage and by so doing take full advantage of the infrastructure without requiring the installation of many station bases. Point to point links were created with planar WiFi antennas; the signal was then replicated with sectorial antennas in all directions. The bandwidth was limited to 256Kb per user due to existing legal restrictions. The tracking system was put activated in 4 homes equipped with a topbox set and a tracking tag in each one. Two mobile phones and 1 user tag were used during the calibration phase.

4 Results and Conclusions

In order to analyze the functioning of the system, the localization and monitoring parts were analyzed separately. The functioning of the passive localization was analyzed first. In order to perform the calibration, the exteriors and the interiors were calibrated. Once all of the WiFi antennas were operative, the exterior calibration involved the use of a mobile terminal with an active GPS and WiFi, which continuously scan the WiFi networks that have been detected at that particular point, and stores the relationship of the WiFi networks with latitude and longitude. The interior calibration was done by using the blueprints of the building located in the mobile. The WiFi networks perform a scan and the user indicates the location on the plans. As the number of access points is low, the precision was set to the room level, where only one position is indicated for each room. Figure 6 shows an image of the mobile calibration system which is performed with the mobile device.

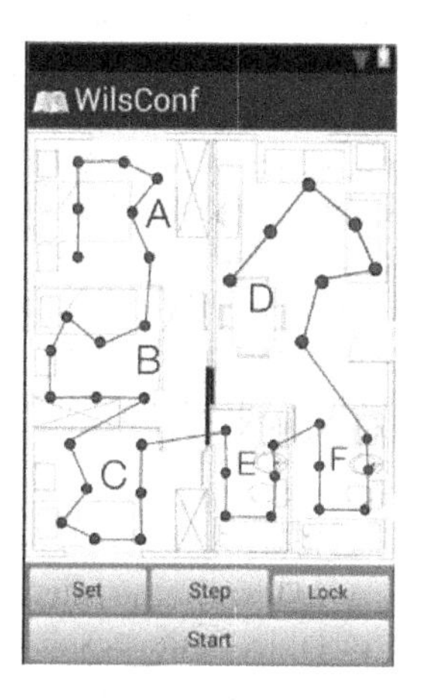

Fig. 6. Screen shot of the calibration application

The precision within one house was analyzed for a one story 85m^2 house with 3 bedrooms, 2 bathrooms and a living room. The calibration included several points within each room with approximately 15 measurements taken for each point, following the diagram shown in Figure 6. The total number of points taken was 40: 7 points per room and living room and 6 points per bathroom. An additional 20 measurements were then randomly taken for each room, living room and bathroom, for a total of 120 measurements that were subsequently classified to determine whether the room had been correctly calculated. The number of networks on each floor varied between 4 and 5. Table 4 shows the number of correct estimates for some of the different techniques that were tested.

Table 4. Rate of accuracy for rooms detected

Technique	Correct estimates
SVM	112
J48	105
MLP	106
LMT	98
KNN	87

An analysis of the battery concluded that it had a life of 37 days, which included refreshing user position every 5 minutes. Battery consumption with the tag in sleep mode is 0.3mA, and 35 mA on standby.

An analysis of the SVM confusion matrix showed that the errors are mainly associated with adjacent bedrooms, which logically explains the results obtained.

The error obtained from the classification process in the town is more difficult to analyze; this is because the GPS already introduces certain errors during the measurement process, and it is complicated to apply the same indoor calibration process to the exterior process because of the time involved. The interquartile range, not related to the GPS provided position, varies between 20 and 50 meters.

In order to analyze the functioning of the monitoring system, it was first necessary to analyze the functioning of the confidence intervals and the classifier. Due to the scarce number of anomalous behavior during this initial phase, it is difficult to analyze the behavior of the system. During the evaluation process the only anomalous cases were manually generated with the specific purpose of analyzing the system. The cases were generated by varying the usual times during which the user had any interaction, and classifying these cases as anomalous. The performance of the different techniques was analyzed using the ROC curves. The same threshold of 0.3 was applied to each of the classifiers. A probability greater than 0.3 was classified as an anomalous situation. As shown in table 5, the result obtained in the area under de ROC curve is greater for LMS than for the other classifiers; nevertheless, it is necessary to increase the number of cases to determine whether the result is satisfactory.

The created system makes it easy to monitor users at a low cost since it is not necessary to install complex hardware, and because the hardware installed has other uses beyond monitoring. The battery life is sufficient enough to suggest its use as an alternative to mobiles, although it would be necessary to find more alternatives to

Table 5. Rate of accuracy for rooms detected

Technique	Correct estimates
SVM	0.84
J48	0.79
MLP	0.88
LMT	0.91
KNN	0.76

prolong battery life, such as installing a motion detector that would only activate the tag when it detects movement. This type of motion detector is already considered in the tag, which includes the required ports. With regard to the monitoring system, it would only be necessary to test it with more real data and load the case memory in the CBR system with real anomalous and non-anomalous cases in order to better analyze the functioning of the system.

Acknowledgments. This work has been supported by International Research Staff Exchange Scheme call FP7-PEOPLE-2012-IRSES, PIRSES-GA-2012-318878.

References

[1] Tusell, F.: Testing for interaction in two-way ANOVA tables with no replication. Computational Statistics & Data Analysis 10(1), 29–45 (1990)
[2] Suryadevara, N.K., Gaddam, A., Rayudu, R.K., Mukhopadhyay, S.C.: Wireless Sensors Network Based Safe Home to Care Elderly People: Behaviour Detection. Procedia Engineering 25, 96–99 (2011)
[3] Noguchi, H., Mori, T., Sato, T.: Construction of network system and the first step of summarization for human daily action in the sensing room. In: Proceedings of the IEEE Workshop on Knowledge Media Networking (KMN 2002) (2002)
[4] Williams, G., Doughty, K., Bradley, D.A.: A systems approach to achieving CarerNet—an integrated and intelligent telecare system. IEEE Trans. Inform. Technol. Biomed. 2(1), 1–9 (1998)
[5] Helal, S., Winkler, B., Lee, K.Y., Ran, L., Giraldo, C., Kuchibhotla, S., Mann, W.: Enabling location-aware pervasive computing applications for the elderly. In: Proceedins of IEEE 1st Conference PerCom 2003 (2003)
[6] Chan, M., Estève, S., Escriba, C., Campo, E.: A review of smart homes—Present state and future challenges 91(1), 55–81 (2008)
[7] Stowe, S., Harding, S.: Telecare, telehealth and telemedicine. European Geriatric Medicine 1(3), 193–197 (2010)
[8] Fourty, N., Guiraud, D., Fraisse, P., Perolle, G., Etxeberria, I., Val, T.: Embedded system used for classifying motor activities of elderly and disabled people. Computers & Industrial Engineering 57(1), 419–432 (2009)
[9] Lesser, V., Atighetchi, M., Benyo, B., Horling, B., Raja, A., Vincent, R., Wagner, T., Xuan, P., Zhang, S.X.Q.: The intelligent home testbed. In: Proceedings of Autonomy Control Software Workshop, vol. 8 (1999)
[10] Pang, S., Ban, T., Kadobayashi, Y., Kasabov, N.: Personalized Mode Transductive Spanning SVM Classification Tree. Information Sciences 181, 2071–2085 (2011)
[11] Corchado, J.M., Bajo, J., Abraham, A.: GERAmI: Improving the delivery of health care in geriatric residences. IEEE Intelligent Systems (2), 19–25
[12] Zato, C., et al.: PANGEA – Platform for Automatic coNstruction of orGanizations of intElligent Agents. In: Omatu, S., Paz Santana, J.F., González, S.R., Molina, J.M., Bernardos, A.M., Rodríguez, J.M.C. (eds.) Distributed Computing and Artificial Intelligence. AISC, vol. 151, pp. 229–240. Springer, Heidelberg (2012)
[13] Landwehr, N., Hall, M., Frank, E.: Logistic Model Trees. Machine Learning 95(1-2), 161–205 (2005)

[14] Tapia, D.I., Alonso, R.S., De Paz, J.F., Zato, C., De la Prieta, F.: International Journal of Artificial Intelligence 6(S11) (2011)
[15] Venturini, V., Carbo, J., Molina, J.M.: Methodological design and comparative evaluation of a MAS providing AmI. Expert Systems with Applications International Journal 39, 10656–10673 (2012)
[16] Sánchez-Pi, N., Carbó, J., Molina, J.M.: A Knowledge-Based System Approach for a Context-Aware System. Knowledge-Based Systems 27, 1–17 (2012)
[17] Gómez, J., Patricio, M.A., García, J., Molina, J.M.: Communication in distributed tracking systems: an ontology-based approach to improve cooperation. Expert Systems 28(4), 288–305 (2011)
[18] Serrano, E., Gómez-Sanz, J.J., Botía, A.A., Pavón, J.: Intelligent data analysis applied to debug complex software systems. Neurocomputing 72(13), 2785–2795 (2009)
[19] Fuentes-Fernández, R., Gómez-Sanz, J.J., Pavón, J.: Understanding the human context in requirements elicitation. Requirements Engineering 15(3), 267–283 (2010)
[20] Fuentes-Fernandez, R., Gomez-Sanz, J.J., Pavon, J.: Model integration in agent-oriented development. International Journal of Agent-Oriented Software Engineering 1(1), 2–27 (2007)
[21] Corchado, J.M., Fyfe, C.: Unsupervised neural method for temperature forecasting. Artificial Intelligence in Engineering 13(4), 351–357 (1999)
[22] Fdez-Riverola, F., Corchado, J.M.: CBR based system for forecasting red tides. Knowledge-Based Systems 16(5), 321–328 (2003)
[23] Tapia, D.I., Abraham, A., Corchado, J.M., Alonso, R.S.: Agents and ambient intelligence: case studies. Journal of Ambient Intelligence and Humanized Computing 1(2), 85–93 (2010)
[24] Corchado, J.M., Lees, B.: Adaptation of cases for case based forecasting with neural network support. In: Soft Computing in Case Based Reasoning, pp. 293–319 (2001)
[25] Corchado, J.M.: Redes Neuronales Artificiales: un enfoque práctico. Servicio de Publicacións da Universidade de Vigo, Vigo (2000)
[26] Bajo, J., Corchado, J.M.: Evaluation and monitoring of the air-sea interaction using a CBR-Agents approach. In: Muñoz-Ávila, H., Ricci, F. (eds.) ICCBR 2005. LNCS (LNAI), vol. 3620, pp. 50–62. Springer, Heidelberg (2005)
[27] Fraile, J.A., Bajo, J., Corchado, J.M., Abraham, A.: Applying wearable solutions in dependent environments. IEEE Transactions on Information Technology in Biomedicine 14(6), 1459–1467 (2011)
[28] Corchado, J.M., De Paz, J.F., Rodríguez, S., Bajo, J.: Model of experts for decision support in the diagnosis of leukemia patients. Artificial Intelligence in Medicine 46(3), 179–200 (2009)
[29] De Paz, J.F., Rodríguez, S., Bajo, J., Corchado, J.M.: Case-based reasoning as a decision support system for cancer diagnosis: A case study. International Journal of Hybrid Intelligent Systems 6(2), 97–110 (2009)
[30] Tapia, D.I., Rodríguez, S., Bajo, J., Corchado, J.M.: FUSION@, a SOA-based multi-agent architecture. In: International Symposium on Distributed Computing and Artificial Intelligence, pp. 99–107 (2008)
[31] Corchado, J.M., Aiken, J.: Hybrid artificial intelligence methods in oceanographic forecast models. IEEE Transactions on Systems, Man, and Cybernetics, Part C: Applications and Reviews 32(4), 307–313 (2002)
[32] Corchado, J.M., Aiken, J., Rees, N.: Artificial intelligence models for oceanographic forecasting. Plymouth Marine Laboratory (2001)

[33] Rodríguez, S., Pérez-Lancho, B., De Paz, J.F., Bajo, J., Corchado, J.M.: Ovamah: Multiagent-based adaptive virtual organizations. In: 12th International Conference on Information Fusion, FUSION 2009, pp. 990–997 (2009)
[34] Tapia, D.I., De Paz, J.F., Rodríguez, S., Bajo, J., Corchado, J.M.: Multi-agent system for security control on industrial environments. International Transactions on System Science and Applications Journal 4(3), 222–226 (2008)
[35] Borrajo, M.L., Baruque, B., Corchado, E., Bajo, J., Corchado, J.M.: Hybrid neural intelligent system to predict business failure in small-to-medium-size enterprises. International Journal of Neural Systems 21(4), 277–296 (2011)
[36] De Paz, J.F., Rodríguez, S., Bajo, J., Corchado, J.M.: Mathematical model for dynamic case-based planning. International Journal of Computer Mathematics 86(10-11), 1719–1730 (2009)
[37] Bajo, J., De Paz, J.F., Rodríguez, S., González, A.: Multi-agent system to monitor oceanic environments. Integrated Computer-Aided Engineering 17(2), 131–144 (2010)
[38] Tapia, D.I., Alonso, R.S., De Paz, J.F., Corchado, J.M.: Introducing a distributed architecture for heterogeneous wireless sensor networks. In: Omatu, S., Rocha, M.P., Bravo, J., Fernández, F., Corchado, E., Bustillo, A., Corchado, J.M. (eds.) IWANN 2009, Part II. LNCS, vol. 5518, pp. 116–123. Springer, Heidelberg (2009)
[39] Rodríguez, S., de Paz, Y., Bajo, J., Corchado, J.M.: Social-based planning model for multiagent systems. Expert Systems with Applications 38(10), 13005–13023 (2011)
[40] Pinzón, C.I., Bajo, J., De Paz, J.F., Corchado, J.M.S.-M.: An adaptive hierarchical distributed multi-agent architecture for blocking malicious SOAP messages within Web Services environments. Expert Systems with Applications 38(5), 5486–5499 (2011)
[41] Corchado, J.M., Bajo, J., De Paz, J.F., Rodríguez, S.: An execution time neural-CBR guidance assistant. Neurocomputing 72(13), 2743–2753 (2009)
[42] Pavon, J., Sansores, C., Gomez-Sanz, J.J.: Modelling and simulation of social systems with INGENIAS. International Journal of Agent-Oriented Software Engineering 2(2), 196–221 (2008)

Context-Aware Module for Social Computing Environments

Gabriel Villarrubia[1], Juan F. De Paz[1], Javier Bajo[2], and Yves Demazeau[3]

[1] BISITE, Universidad de Salamanca, Salamanca, Spain
{gvg,fcofds}@usal.es
[2] DIA, Universidad Politécnica de Madrid, Madrid, Spain
jbajo@fi.upm.es
[3] CNRS, LIG, Grenoble, France
Yves.Demazeau@imag.fr

Abstract. The continuous evolution of the information and telecommunication technologies has led to new forms of social interaction, including social networks. Social interaction is a new paradigm that studies the use of information technologies with social purposes. Social computing envisions a new kind of computation where humans and machines collaborate to compute and resolve a problem. In this paper we present a context-aware module for the PANGEA architecture that incorporates contextual information to enrich the social knowledge representation.

Keywords: Multi-agent systems, Human-agent societies, Context-Aware Computing.

1 Introduction

During recent years social computational solutions have emerged to provide new ways for interaction and communication. Some examples are Amazon, where the humans contribute to the computation including their opinions and recommendations about the products, Captcha [25], where both humans and computers collaborate to provide an efficient authentication system, etc. Social computing is a new computational model where human and computers collaborate to improve social relationships using computer science [26], [4], [7], [15], [7], [9]. For Wang et al. [26] Social Computing is the computational facilitation of social studies and human social dynamics as well as the design and use of ICT technologies that consider social context. In this sense, it is important to define new mechanisms to include contextual information in the social computing model. For Robertson et al. [17] social computing requires and effective combination of computational and human resources: On the one hand, humans bring their competences, knowledge and skills, together with their networks of social relationships and their understanding of social structures. On the other hand, ICT can search for and deliver relevant information. Humans can then use this information within their contexts to achieve their goals and, eventually, to improve the overall environment in which they live.

C. Ramos et al. (eds.), *Ambient Intelligence - Software and Applications*,
Advances in Intelligent Systems and Computing 291,
DOI: 10.1007/978-3-319-07596-9_20, © Springer International Publishing Switzerland 2014

One of the open challenges for Social Computing is to provide more realistic ways to improve social behaviors and relationships using computer science. The existing solutions have focused on theoretical underprintings, technological infrastructure and applications. However, it is necessary to capture contextual information to enrich the social model, thus providing more realistic computational tools for the cooperation between humans and computers. In this paper, we present an extension of the PANGEA [29] architecture a multiagent architecture based on virtual organizations [2], [3], that incorporates a context-aware computing model [27], [23], [21] to obtain contextual information. In this paper, we extend the PANGEA architecture and we define all the infrastructure components, both sensor networks and computing, especially at a hardware level. We define a broker that interacts with the sensing technologies and a set of adapters that normalize the data. The broker communicates with the rest of the platform by means of adapters. We design a new model to integrate the JDL information fusion model within the virtual organization-based multiagent architecture [4], [18], [23], [19]. Particularly, we focus on designing new algorithms for mixtures of experts specialized in fusing information obtained from wireless sensor networks.

The rest of the paper is organized as follows: section 2 revises the related work. Section 3 presents the proposed model. Finally, in section 4 the preliminary conclusions obtained are presented.

2 Related Work

Recent tendencies have led to the social computing paradigm of designing social systems. One of the challenges to be addressed to obtain an extended model with context-aware computing abilities is the procurement of effective management architectures for WSNs. Until now, WSNs and their applications have been developed without considering a management solution that can dynamically adapt to both the changes that occur in the environment, and to user needs. Some approaches as the MANNA management architecture for WSNs propose the functional, information, and physical management architectures, that take into account the specific characteristics of this type of network [12]. However, this architecture does not take into account either adaptive and organizational aspects, or intelligent information fusion (IF). Lim *et al.* [12][13] propose a sensor grid architecture, called the scalable proxy-based architecture for sensor grid (SPRING), to address these design issues [12]. However, the architecture is focused on a sensor grid design and not on exploitation. H-WSNMS uses the concept of a virtual command set, H-WSNMS, to facilitate management functions for specific WSN applications from the individual WSN platforms [26], but does not take IF algorithms into account and is not designed on the basis of organizational aspects. MARWIS is a management architecture for heterogeneous wireless sensor networks (WSNs). It supports common management tasks such as monitoring, (re-)configuration, and updating program code in a WSN [21]. MARWIS, however, does not take organizational aspects into account and does not fuse information. Yu *et al.* [26] propose a lightweight middleware system that supports WSNs to handle real-time network management using a hierarchical framework [27]. Although they take organizational aspects into account, they do not consider IF algorithms and user services. G-Sense [15] is an architecture that integrates mobile and static wireless sensor networks in support of location-based services, participatory sensing, and human-centric sensing applications. It does not, however, take organizational aspects into account, nor does it include IF technologies. Nowadays it is possible to find different proposals for architectures that manage wireless sensor networks [7][14][4][12]; however, most of them are designed for specific environments or specific purposes and none of them combines organizational aspects, IF techniques, advanced storage mechanisms and open integration design.

Although significant progress has been made in the development of architectures to manage wireless sensor networks, at present there is no single open platform that efficiently integrates heterogeneous WSNs, and provides both intelligent IF techniques and intelligent services. Therefore, there is no platform in the market that facilitates the communication and integration of the wide variety of existing sensors, providing intelligent IF facilities, intelligent management of user services. The proposed Virtual Organization (VO) of multiagent architecture is based on the social computing paradigm and will provide intelligence to the platform with adaptation to the needs of the application problem, while the cloud environment will ensure the availability of the required resources at all times.

3 PANGEA Architecture

PANGEA (Platform for Automatic coNstruction of orGanizations of intElligents Agents) [29] is an agent platform to develop open multi-agent systems; it can manages roles, norms, organizations and suborganizations that facilitate the inclusion of organizational aspects. The services offered by the agents are included independently from the agent, facilitating their flexibility and adaption. PANGEA incorporates a CBR-BDI reasoning mechanism available for the agents. The basic agent types defined in PANGEA can be seen in Figure 1, they are:

- OrganizationManager: the agent responsible for the actual management of organizations and suborganizations. It is responsible for verifying the entry and exit of agents, and for assigning roles. To carry out these tasks, it works with the OrganizationAgent, which is a specialized version of this agent.
- InformationAgent: the agent responsible for accessing the database containing all pertinent system information.
- ServiceAgent: the agent responsible for recording and controlling the operation of services offered by the agents. It works as the Directory Facilitator defined in the FIPA standard.
- NormAgent: the agent that ensures compliance with all the refined norms in the organization.

- CommunicationAgent: the agent responsible for controlling communication among agents, and for recording the interaction between agents and organizations.
- Sniffer: manages the message history and filters information by controlling communication initiated by queries.
- DiscoveryAgent: implements an intelligent mechanism to discover services.
- MonitorAgent: interacts with the platform to show the information to the end user.

PANGEA is a service-oriented platform that can take maximum advantage of the distribution of resources. To this end, all services are implemented as Web Services. This makes it possible for the platform to include both a service provider agent and a consumer agent, thus emulating a client-server architecture. The provider agent (a general agent that provide a service) knows how to contact the web service, the rest of the agents know how to contact with the provider agent due to their communication with the ServiceAgent, which contains this informacion about services.

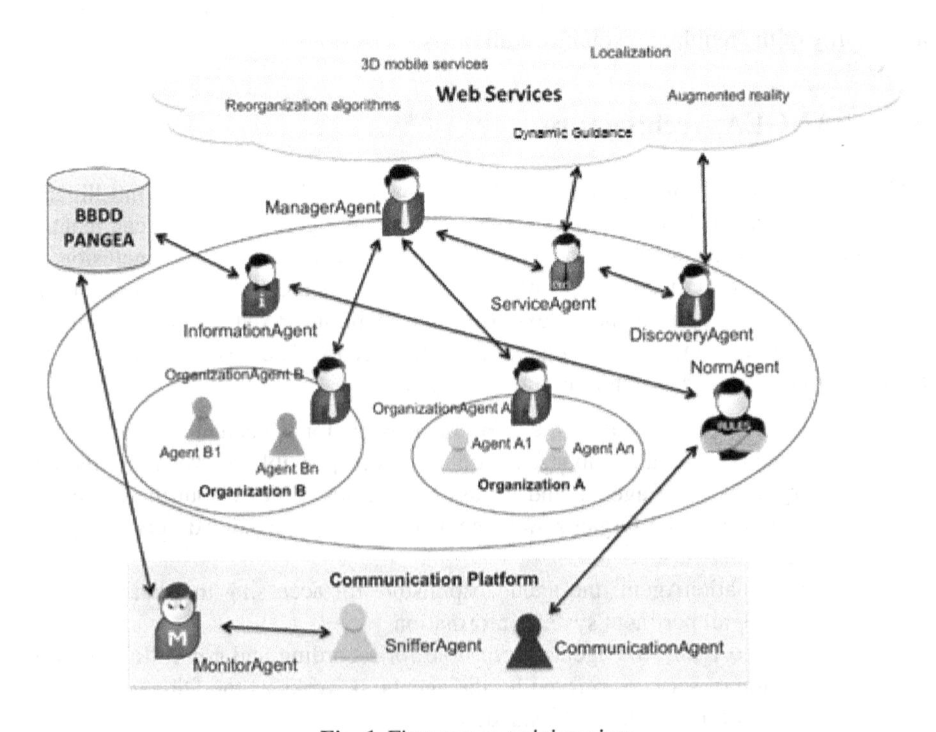

Fig. 1. First-person training view

3.1 Extensions for Context-Aware Computing

The extended model for the PANGEA architecture allows obtaining and managing contextual information that can provide an added value to design social computing models and obtain immersion at the MAS level. Contextual information can enrich the humans' information providing more realistic data about the humans' situation, including human actions and behaviors in a given society. Context-aware systems manage information that characterizes an individual and her environment. These systems require sensor networks to capture the context information and intelligent systems that can manage the information efficiently.

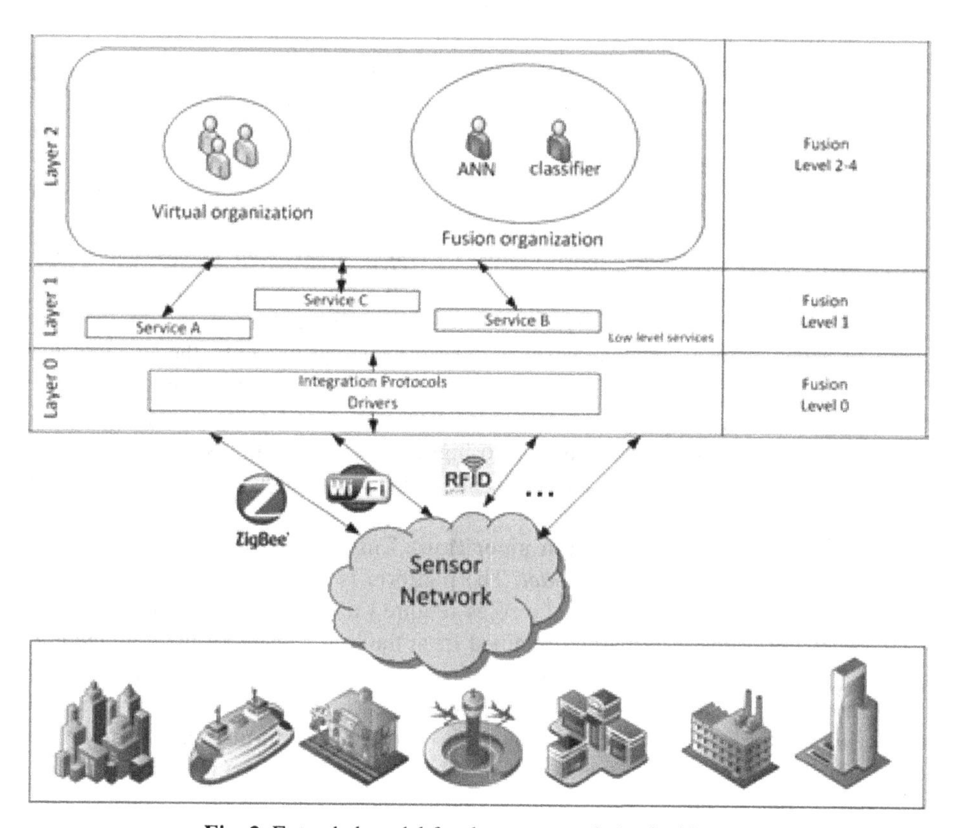

Fig. 2. Extended model for the sensor and physical layers

The proposed extended model is deployed into a layered architecture as shown in Figure 2. As we can see, the extended model is composed of different layers associated with the different functional blocks. The fusion levels of information are distributed along the different layers that can be found in the description of the JDL fusion model [1][17]. As defined in JDL, existing levels of data fusion from 0 to 6 are Data Assessment, Object Assessment, Situation Assessment, Impact Assessment, and Process Refinement. These levels are distributed in the different layers of the architecture shown in Figure 2.

The following section describes the components and main features of the architecture:

- **Layer 0. Sensing/performance technologies.** Layer 0 of the platform is a broker that defines communication with sensor networks of different natures (Wi-Fi, ZigBee, Bluetooth, etc.), and obtains the raw data from sensor networks. This process of acquiring raw data from sensor networks is associated with *Level 0 - Data Assessment*. The main novelty of this layer is the ability to provide the platform and the upper layers with openness regarding the connection to sensor networks of different natures. It thus ensures that upper layers of the architecture have access to information and are able to perform data fusion at different levels.
- **Layer 1. Low-level services.** Given the information exchanged with the environment through layer 0 as described above, the existing functional requirements and a set of low-level services will now be defined; specifically those that depend on the types of networks and technologies integrated into every deployment. After obtaining the raw data, a gateway is provided, defined through adapters that allow the information received to be standardized. The data processing corresponds to *Level 1 - Object Assessment* as indicated by the JDL classification shown above. In this first stage, the platform provides services such as filtering of signals, normalization services or other treatment services at the basic level signals. These services are provided by the adapters and is associated with algorithms that perform initial treatment of the data, so that these data can be presented to higher layers in a more homogenized way. Each of these services expose an API to higher layers that allows interaction with each low-level service, and thus, with the underlying sensing/performance technologies.
- **Layer 2. Information fusion algorithms.** This layer includes le*vels 2 to 4 of IF displayed on the JDL model*. The platform is structured as a VO of MAS. Each organization includes the roles required to facilitate an intelligent management of the information obtained from the lower levels of the architecture. The MAS incorporates agents specifically designed to interact with low-level services. In addition, we introduce the design of intelligent agents specialized in IF. For this purpose, roles that allow merging information automatically through supervised learning and previous training have been included.

4 Conclusions

This paper has presented an extended model for social computing for the PANGEA architecture, aimed at extending the concept of social computing to obtain immersion at the MAS level. Social computing has gained relevance during recent years, trying to combine sociology and computer sciences to design social systems. The extended model proposed in this paper is currently being detailed and evaluated using location-aware systems. A location-aware system can notably help to obtain real-time information that can be used for social computing purposes in different applications that can make use of location data to create social machines. Some examples can be the prediction of social dynamics, design of activities, urban architectural design, management of emergency situations or other several social behaviors that can be analyzed and supported by computational technologies. Particularly, in future work, we want to focus on a case study to design an agent-based social simulation model for work environment and to introduce new variables in the simulation model, such as individual and group behavior obtained from the location of the participants. The agent-based social simulation model will be an extension of a previous model aimed at emulate human behaviours in a work environment to predict the labour integration of handicapped people.

The inclusion of a context-aware module in the PANGEA architecture can help us to introduce and analyse contextual information and patterns related to interaction and collaboration behaviours that usually are hidden in real societies. More specifically, the use of location-aware techniques can help us to detect friendship or other collaboration events that cannot be regulated in current agent-based social simulation models. The use of fusion will allow the implementation of specific experts, signal agents to process and filter signal data. One of the main advantages of this implementation is the high adaptability of the platform to dynamically incorporate new agents (experts, mixtures, filtering algorithms, etc.). Thus, a fusion agent will make use of the information provided by two or more expert agents to generate high level information services. The IF process requires communication among the different layers of the architecture that will be implemented through message passing protocols.

Acknowledgments. Work partially supported by the Spanish Government through the project iHAS (grant TIN2012-36586-C03).

References

1. Abras, S., Ploix, S., Pesty, S., Jacomino, M.: A multi-agent home automation system for power management. LNEE, vol. 15, pp. 59–68 (2008)
2. Artikis, A.: Dynamic Protocols for Open Agent Systems. In: Decker, S., Sierra, C. (eds.) Proc. of 8th Int. Conf. on Autonomous Agents and Multiagent Systems, AAMAS 2009. International Foundation for Autonomous Agents and Multiagent SystemsMay, pp. 10–15 (2009)
3. Argente, E., Botti, V., Carrascosa, C., Giret, A., Julian, V., Rebollo, M.: An Abstract Architecture for Virtual Organizations: The THOMAS approach. Knowledge and Information Systems 29, 379–403 (2011)
4. Bajo, J., Corchado, J.M.: Evaluation and monitoring of the air-sea interaction using a CBR-Agents approach. Case-Based Reasoning Research and Development, 50–62 (2005)
5. Carneiro, D., Castillo, J., Novais, P., Fernández-Caballero, A., Neves, J.: Multimodal Behavioural Analysis for Non-invasive Stress Detection. Expert Systems With Applications 39(18), 13376–13389 (2012) ISSN: 0957-4174
6. Dimeas, A.L., Hatziargyriou, N.D.: Operation of a Multiagent System for Microgrid Control. IEEE Transactions on Power Systems 20(3), 1447–1455 (2005)

7. Fraile, J.A., Bajo, J., Corchado, J.M., Abraham, A.: Applying wearable solutions in dependent environments. IEEE Transactions on Information Technology in Biomedicine 14(6), 1459–1467 (2011)
8. Fuentes-Fernández, R., Gómez-Sanz, J.J., Pavón, J.: Understanding the human context in requirements elicitation. Requirements engineering 15(3), 267–283 (2010)
9. Fuentes-Fernandez, R., Gomez-Sanz, J.J., Pavon, J.: Model integration in agent-oriented development. International Journal of Agent-Oriented Software Engineering 1(1), 2–27 (2007)
10. Gómez-Romero, J., Serrano, M.A., Patricio, M.A., García, J., Molina, J.M.: Context-based Scene Recognition from Visual Data in Smart Homes: An Information Fusion approach. ACM/Springer Journal of Personal and Ubiquitous Computing. Special Issue on Sensor-driven Computing and Applications for Ambient Intelligence 16(7), 835–857 (2012)
11. Haibo, L., Fang, Z.: Design and implementation of wireless sensor network management systems based on WEBGIS. Journal of Theoretical and Applied Information Technology 49(2), 792–797 (2013)
12. Karim, L., Mahmoud, Q.H., Nasser, N., Khan, N.: LRSA: A multi-component Wireless Sensor Network management framework. In: 2012 IEEE Global Communications Conference (GLOBECOM), pp. 683–688 (2012)
13. Lim, H.B., Iqbal, M., Wang, W., Yao, Y.: The National Weather Sensor Grid: a large-scale cyber-sensor infrastructure for environmental monitoring. IJSNet 7(1/2), 19–36 (2010)
14. Liu, Y., Seet, B.C., Al-Anbuky., A.: An Ontology-Based Context Model for Wireless Sensor Network (WSN) Management in the Internet of Things. J. Sens. Actuator Netw. 2(4), 653–674 (2013)
15. Pavon, J., Sansores, C., Gomez-Sanz, J.J.: Modelling and simulation of social systems with INGENIAS. International Journal of Agent-Oriented Software Engineering 2(2), 196–221 (2008)
16. Perez, A.J., Labrador, M.A., Barbeau, S.J.: G-Sense: a scalable architecture for global sensing and monitoring. IEEE Network 24(4), 57–64 (2010)
17. Robertson, D., Giunchiglia, F.: Programming the social computer. Phil. Trans. R. Soc. A. 371, 20120379
18. Rodríguez, S., Pérez-Lancho, B., De Paz, J.F., Bajo, J., Corchado, J.M.: Ovamah: Multiagent-based adaptive virtual organizations. In: 12th International Conference on Information Fusion, FUSION 2009, pp. 990–997 (2009)
19. Rodríguez, S., de Paz, Y., Bajo, J., Corchado, J.: Social-based planning model for multiagent systems. Expert Systems with Applications 38(10), 13005–13023 (2011)
20. Schuler, D.: Social computing. Communications of the ACM Volume 37(1), 28–29 (1994)
21. Tapia, D.I., Alonso, R.S., De Paz, J.F., Corchado, J.: Introducing a distributed architecture for heterogeneous wireless sensor networks. In: Omatu, S., Rocha, M.P., Bravo, J., Fernández, F., Corchado, E., Bustillo, A., Corchado, J.M. (eds.) IWANN 2009, Part II. LNCS, vol. 5518, pp. 116–123. Springer, Heidelberg (2009)
22. Tapia, D.I., Abraham, A., Corchado, J.M., Alonso, R.: Agents and ambient intelligence: case studies. Journal of Ambient Intelligence and Humanized Computing 1(2), 85–93 (2010)
23. Tapia, D.I., Rodríguez, S., Bajo, J., Corchado, J.M.: FUSION@, a SOA-based multi-agent architecture. In: International Symposium on Distributed Computing and Artificial Intelligence, DCAI 2008, pp. 99–107 (2008)

24. Tapia, D.I., Alonso, R.S., García, Ó., de la Prieta, F., Pérez-Lancho, B.: Cloud-IO: Cloud computing platform for the fast deployment of services over wireless sensor networks. In: Uden, L., Herrera, F., Bajo, J., Corchado, J.M. (eds.) 7th International Conference on KMO. Advances in Intelligent Systems and Computing, vol. 172, pp. 493–504. Springer, Heidelberg (2013)
25. Von Ahn, L., Dabbish, L.: Designing games with a purpose. Communications of the ACM 51(8), 58–67 (2008)
26. Wang, F.Y., Carley, K.M., Zeng, D., Mao, W.: Social Computing: From Social Informatics to Social Intelligence. IEEE Intelligent Systems 22(2), 79–83 (2007)
27. Venturini, V., Carbo, J., Molina, J.M.: Methodological design and comparative evaluation of a MAS providing AmI. Expert Systems with Applications 39, 10656–10673 (2012)
28. Wong, J.K.W., Li, H., Wang, S.W.: Intelligent building research: a review. Automation in Construction 14(1), 143–159 (2005)
29. Zato, C., Villarrubia, G., Sánchez, A., Barri, I., Rubión, E., Fernández, A., Rebate, C., Cabo, J.A., Álamos, T., Sanz, J., Seco, J., Bajo, J.J.: MCorchado. PANGEA - Platform for Automatic coNstruction of orGanizations of intElligent Agents. In: Distributed 9th International Conference Computing and Artificial Intelligence (2012)

Wireless Multisensory Interaction in an Intelligent Rehabilitation Environment

Miguel Oliver, José Pascual Molina, Francisco Montero, Pascual González, and Antonio Fernández-Caballero

[1] Instituto de Investigación en Informática de Albacete (I3A), 02071-Albacete, Spain
[2] Universidad de Castilla-La Mancha, Departamento de Sistemas Informáticos, 02071-Albacete, Spain
oliver@dsi.uclm.es

Abstract. Today, the population is aging, and this is becoming a problem for current health systems, as each day it has to invest more money in treating the elderly. Rehabilitation of elderly patients with physical disabilities is one of these problems that everyday incur greater overhead to health care. This paper provides a gerontechnology-based solution by proposing a multisensory system for rehabilitation in an intelligent environment. The proposal enables helping needed people and thus reducing the cost of health care.

Keywords: Gerontechnology, Rehabilitation, Intelligent environments, Multisensory interaction, Wireless sensor networks.

1 Introduction

Nowadays, the population is aging, and this is becoming a problem for current health systems, as each day it has to invest more money in treating the elderly [1], [2]. Rehabilitation is one of these problems that everyday incur greater overhead to health care.

Gerontechnology is an interdisciplinary field of scientific research in which technology is directed towards the aspirations and opportunities for the older persons [3]. Gerontechnology aims at good health, full social participation and independent living up to a high age, be it research, development or design of products and services to increase the quality of life. The development of gerontechnology-based support systems for rehabilitation follows two distinct paths. On one hand, some approaches make use of different specialized hardware to aid the elderly patient in his/her recovery [4]. Generally, these solutions often require a large outlay of money, making access to these products not universal. On the other hand, some other systems propose the use of depth sensors like Kinect to develop a rehabilitation system [5]. This makes the system affordable to the general public, but the interaction is not completely satisfactory in some cases.

In this paper we propose to combine the advantages of both types of systems discussed above. For this, a Kinect sensor is used to pick up the movements made by the elderly in his/her rehabilitation process. In addition, the Wiimote actuator is used to provide haptic sensations to the patient. Thus, the resulting communication is enriched and the elderly is informed about the complete rehabilitation process.

C. Ramos et al. (eds.), *Ambient Intelligence - Software and Applications*,
Advances in Intelligent Systems and Computing 291,
DOI: 10.1007/978-3-319-07596-9_21,

In the next section some rehabilitation systems will be shown, these will help us to create a system of rehabilitation for elderly people that will be displayed in the third section. Finally the conclusions obtained during the development of the rehabilitation system are discussed, which lead us to say that the project developed allows the rehabilitation of elderly people from home, which means saving time and money for both the patient and rehabilitation center.

2 Some Current Rehabilitation Systems

In the literature there are several examples of the use of depth sensors such as the Kinect sensor as part of support systems in rehabilitation [6]. In a recent paper [5] the possibility of using the Kinect sensor for the rehabilitation of patients with motor impairments is studied. The system is used by two young participants, resulting in improved patient motivation in rehabilitation and improving stretching exercise performance. Another paper [7] contains a study which tests the Kinect sensor as a substitute for classic assisted rehabilitation, offering the user the possibility of rehabilitation at home. The result of the study shows that the reliability in gesture recognition is between 88.0% and 92.2%. Moreover, the study proves that the participants find more fun and ease to use in the Kinect than in the conventional system. Another development and evaluation of a rehabilitation system based on Kinect has been presented [8]. The proposed system is composed of two modules. The first consists of a set of support stretching exercises for physical rehabilitation, and the second, in a data analyzer which detects posture and wrong actions on the user. Also, a comparison of low cost sensor Kinect and sensor motion capture OptiTrac V100:R2 of the company NaturalPoint has been presented [9]. The comparison raises that the Kinect sensor provides an acceptable performance that is competitive with the sensor OptiTrac V100:R2, but with a much lower price. The proposal allows universal accessibility to computerized rehabilitation treatments. In addition, other papers (e.g. [10], [11], [12]) propose the use of devices such as the Kinect sensor and the development of games to help patients in their rehabilitation.

Moreover, to date, some physical systems have been implemented for the rehabilitation of patients. *VirtualRehab* [13], developed by company Virtualware Group, is a product for the rehabilitation of patients with any degree of physical disability. The main feature of its software is that it deals with the issue of rehabilitation as a game. This ensures the software to have a playful component, which induces the user to feel more comfortable during the treatment. At present, the system has 9 games/exercises that the physiotherapist assigns to a patient depending on his/her disability.

Teki [14] is a project of the Basque Health Service (Osakidetza). The project seeks to improve the quality of life of chronic patients, providing a tool to monitor clinical status and to facilitate greater communication between the patient and the specialist, thereby improving the care of the user at home by using new information technologies.*Teki* allows the collection of clinical data using medical devices, the response to symptom questionnaires and the recording of self-administration of medication, which allows to remotely monitor the patient.

Toyra [15] is a rehabilitation product developed by the company Indra. It is aimed to the rehabilitation of the upper body, and for it, makes use of Kinect sensor and a set of sensors that have to be attached to the body. These sensors measure the movement of the human body and transmit it to the computer, which adds this data to the information collected by the Kinect, to form a set of more accurate data. The system is divided into *Assisted Toyra*, which is aimed for use in specialized centers with the supervision of a physiotherapist, and *Independent Toyra*, with is aimed at the rehabilitation of the patient at home.

Brontes Processing [16] is a company dedicated to developing games which base their interaction on webcams for personal computers. At present, the company offers two products for use in rehabilitation: *SeeMe* and *Home rehabilitation*. The *SeeMe* system is used for the rehabilitation of patients in specialized centers, requiring the support of specialized doctors for proper operation. Now, *Home rehabilitation* unlike the previous one, is used for the rehabilitation of patients at home.

Reflexion [17] is a system developed by the medical West Health Institute. Its main purpose is to maintain the interest of the patient throughout the treatment and to inform the therapist about the correction in the conduct of the stretching exercise. The specialist selects and adapts the rehabilitation exercises to each patient, who performs the exercises at home or at a place suitable for use by a computer and a Kinect sensor.

Lastly, *KineLabs* [18] is a project at the Hong Kong Polytechnic University for help in the rehabilitation of the elderly and people with physical disabilities. The system currently contains three stretching exercises which are aimed at coordination of the upper limbs, coordination of the lower limbs and trunk balance training.

There are two disadvantages in the systems proposed in this section. The first disadvantage is that these systems are based on predetermined exercises which are assigned to the patients, this means that the exercises are generic and cannot adjust to the peculiarities of each of these patients. The second disadvantage of these systems is the use of visual and auditory stimuli, for patients with visual and audible deficiencies, this can make, that the rehabilitation occurs incorrectly. Considering these disadvantages, we will create the system of rehabilitation for elderly patients.

3 A New Proposal of a Rehabilitation System

Our system is being developed to assist in the rehabilitation of elderly patients with physical disabilities. To accomplish this, the therapist establishes a set of stretching exercises that the patients have to do. The system is responsible of handling the rehabilitation exercises and evaluating their adequacy, taking as example the stretching exercises provided by the monitor. In addition, the system provides feedback to the user in relation to the correctness of the exercise performed. During the system development, we considered the option of providing an audible feedback to the user, but this stimulus can be confusing and annoying in the moment that there are several rehabilitation patients in a same room, as it usually happens in rehabilitation groups. Also the elderly people, the common user of this system, may have problems with this stimulus due to their hearing problems. Because of these reasons, we decided to add an audio feedback to the system, but augmented with another sensory channel. Therefore, we performed a study on haptics to supplement the auditory canal. As a result, we reached the idea of using vibration motors to enhance the communication between the user and the system. More specifically, we chose to use a Wiimote game controller used on the Wii console.

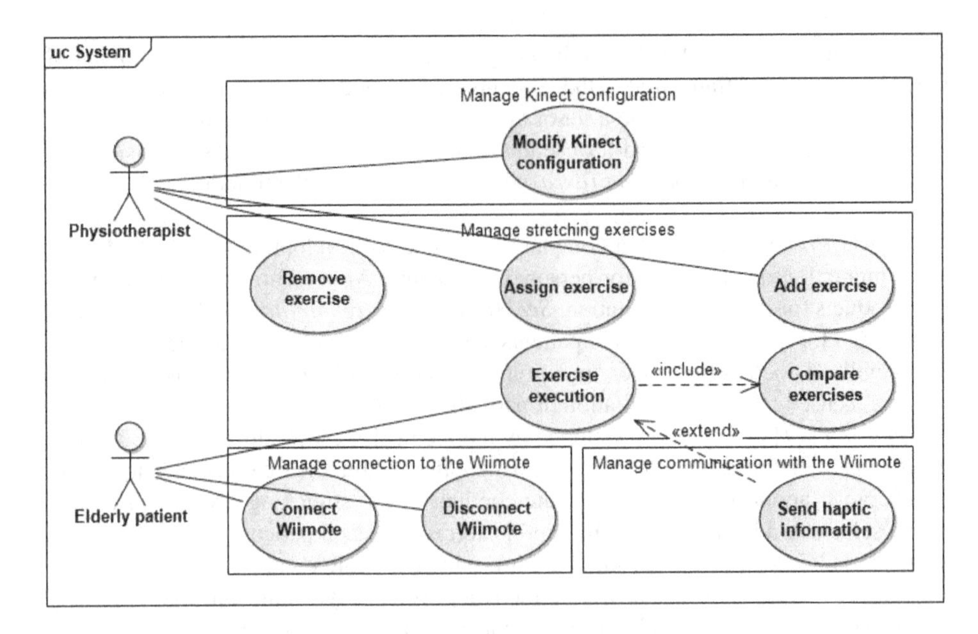

Fig. 1. Use case diagram of the rehabilitation system

Fig. 2. Diagram of hardware components of the system

Fig. 1 shows a use case diagram of the developed system. It describes how the system is composed of two users, the therapist and the elderly patient, as well as the tasks performed by each of them. The most important tasks are described later on.

Fig. 2 shows the hardware devices that compose the rehabilitation system. The heart of the system is the computer. It is responsible for processing the signals of all devices and for running the developed software. The Kinect device is connected to the PC via USB and is responsible for collecting and preprocessing the images of the therapist and the elderly patient. The Wiimote device is connected to the PC using a Bluetooth connection and is responsible for providing haptic feedback to the patient. This way the elderly patient determines if he/she is performing the stretching exercise correctly. The arrows in the figure indicate the flow of information.

In first place, the developed system looks for a Wiimote device to provide haptic feedback to the user. If there is at least one paired device, using Bluetooth, the system recognizes it and connect to it. If there is no Wiimote connected to the computer, the rehabilitation system warns about this fact and asks the user if he/she wants to search again a device or to run without haptic feedback. Fig. 3 shows the main interface of

Fig. 3. Elderly person performing a shoulder rehabilitation exercise

the program and an elderly patient performing an stretching exercise proposed by a physiotherapist. The center-left of the screen shows the image captured in real-time by the Kinect sensor. On the right side of the interface there is a tab control that offers four different tabs used both by the elderly patient and the physical therapist to navigate the application and to interact with it.

Next, the operation of the system is described according to the functionality required by each system role.

3.1 Physiotherapist

The role of the physiotherapist in the system is relatively simple. He/she adds the stretching exercises, assigns these rehabilitation exercises to elderly patients, and adjusts the characteristics of the system to adapt it to the environment where the Kinect device is located.

1. *Add exercise*. To add new stretching exercises to the system, the therapist must select "Add exercise". Once done, a similar interface to that shown in Fig. 4 appears. The therapist assigns a name to the rehabilitation exercise and he/she adopts a position with his/her body that fits the desired rehabilitation. For this part, he/she makes use of the image displayed in the main window which collects the image in real-time of the therapist as well as his/her skeleton. Once the correct posture for rehabilitation is adopted and the Kinect sensor has recognized it, he/she adds the position to the system database.
2. *Assign exercise*. To assign stretching exercises to each patient, the therapist must select "Assign exercise". This will show all the rehabilitation exercises stored in system and provides the option to enable or disable this exercise for a particular user. So the elderly patient can choose only those exercises that the specialty was assigned him/her.

Fig. 4. Physiotherapist creating a shoulder rehabilitation exercise

3. *Modify Kinect configuration*. Another task is to establish the correct parameters of the Kinect sensor to make it work in a proper way. This is achieved by selecting the "Configuration". The configuration options allow to adjust many of the properties of the sensor. Some of the most important features are: the resolution of image capture, color and depth; the recognition of the user standing (Normal mode) or sitting (Near mode); the determination of the recognized user in case there are more users captured by the camera; the inclination of sensor. This allows the Kinect to perfectly conform to the medium in which it is located, thereby improving the resulting interaction with the system.

3.2 Elderly Patient

The role of the elderly patient in the system is to perform stretching exercises that the physiotherapist has previously established. To do this, he/she just selects one of the available exercises and does it as well as possible.

Exercise execution enables the user to select one of the stretching exercises assigned by the physiotherapist. The interface shows all available positions which are composed of a descriptive text and an image showing the therapist performing the rehabilitation exercise. The user just double clicks on the desired exercise to start. Once the exercise has been selected, an interface similar to that of Fig. 3 is automatically displayed. The elderly patient aided by the real-time image captured by the sensor should adopts a similar posture to the sample. To ensure that the patient's posture is similar to the physiotherapist's one, the bottom bar shows the degree of concordance between the two positions. This encourages the users to perform better exercises and this helps in their rehabilitation. Fig. 3 shows a patient performing an stretching exercise, the Wiimote controller, which can be held by hand or stored in his pocket, vibrate when the elderly person performing the rehabilitation exercise correctly. The vibration frequency and intensity depends on the correction in the development of exercise, becoming maximal when the exercise is completely well performed. So, an additional help identifies how he/she is conducting an exercise, helping him/her to continue the rehabilitation.

4 Discussion and Conclusions

Most systems proposed in the literature make use of rehabilitation games. The main objective is to involve the user in the process of rehabilitation and to prevent the treatment abandonment. Our system follows this principle to make more pleasant the elderly patient rehabilitation. In contrast, the previously proposed systems make use of default stretching exercises, which are assigned to the patient in accordance with the necessary rehabilitation. This can make that the specific requirements of a elderly patient are not covered. Our system improves this deficiency by ensuring that the stretching exercises are specifically tailored to the elderly patient and fully meet his/her needs. Visual and audio interaction are used in these systems. But this also causes a problem in rehabilitation of patients with audible and visual impairment, which also coincides with the common user who uses to require rehabilitation, the elderly people. Our system has added haptic interaction to help elderly patients in identifying how they are performing the rehabilitation exercises.

The distinguishing characteristics of our system allow the rehabilitation of a general public and their ease of use allows access to people with no computer skills. Also, they reduce the rehabilitation cost of both the elderly patient and the rehabilitation center, as they enable the rehabilitation of elderly patients at their own home, thereby saving the constant supervision necessary in common rehabilitation. At this moment, the systems allows the execution of rehabilitation exercises which do not require high precision, e.g. arms and legs. However we are working in the improvement of the system to perform precision exercises, such as finger rehabilitation.

Acknowledgements. This work was partially supported by Spanish Ministerio de Economía y Competitividad / FEDER under TIN2010-20845-C03-01 and TIN2012-34003 grants. We would also like to thank RedAUTI (512RT0461) for the comments and suggestions which allowed to greatly improve the quality of the paper.

References

1. Costa, A., Castillo, J.C., Novais, P., Fernández-Caballero, A., Simoes, R.: Sensor-driven agenda for intelligent home care of the elderly. Expert Systems with Applications 39(15), 12192–12204 (2012)
2. Castillo, J.A., Carneiro, D., Serrano-Cuerda, J., Novais, P., Fernández-Caballero, A., Neves, J.: A multi-modal approach for activity classification and fall detection. International Journal of Systems Science 45(4), 810–824 (2014)
3. Harrington, T.L., Harrington, M.K.: Gerontechnology: Why and How. Shaker Publishing BV (2000)

4. Schönauer, C., Pintaric, T., Kaufmann, H., Jansen-Kosterink, S., Vollenbroek-Hutten, M.: Chronic pain rehabilitation with a serious game using multimodal input. In: Proceedings of the 2011 International Conference on Virtual Rehabilitation, pp. 1–8 (2011)
5. Chang, Y., Chen, S., Huang, J.: A Kinect-based system for physical rehabilitation: A pilot study for young adults with motor disabilities. Research in Developmental Disabilities 32(6), 2566–2570 (2011)
6. Fuentes, J.A., Oliver, M., Montero, F., Fernández-Caballero, A., Fernández, M.A.: Towards usability evaluation of multimodal assistive technologies using RGB-D sensors. In: Ferrández Vicente, J.M., Álvarez Sánchez, J.R., de la Paz López, F., Toledo Moreo, F. J. (eds.) IWINAC 2013, Part II. LNCS, vol. 7931, pp. 210–219. Springer, Heidelberg (2013)
7. Davaasambuu, E., Chiang, C., Chiang, J., Chen, Y., Bilgee, S.: A Microsoft Kinect based virtual rehabilitation system. In: Proceedings of the 5th International Conference on Frontiers of Information Technology, Applications and Tools, pp. 44–50 (2012)
8. Freitas, D., Da Gama, A., Figueiredo, L., Chaves, T., Marques-Oliveira, D., Teichrieb, V., Araújo, C.: Development and evaluation of a Kinect based motor rehabilitation game. In: Proceedings of Simposio Brasileiro de Jogos e Entretenimento Digital, pp. 144–153 (2012)
9. Chang, C., Lange, B., Zhang, M., Koenig, S., Requejo, P., Somboon, N., Sawchuk, A., Rizzo, A.: Towards pervasive physical rehabilitation using Microsoft Kinect. In: Proceedings of the 6th International Conference on Pervasive Computing Technologies for Healthcare, pp. 1–4 (2012)
10. Lange, B., Chang, C., Suma, E., Newman, B., Rizzo, A., Bolas, M.: Development and evaluation of low cost game-based balance rehabilitation tool using the Microsoft Kinect sensor. In: Proceedings of the 33rd Annual International Conference of the IEEE Engineering in Medicine and Biology Society, pp. 1831–1834 (2011)
11. Pirovano, M., Mainetti, R., Baud-Bovy, G., Lanzi, P., Borghese, N.: Self-adaptive games for rehabilitation at home. In: Proceedings of the IEEE Conference on Computational Intelligence and Games, pp. 179–186 (2012)
12. Da Gama, A., Chaves, T., Figueiredo, L., Teichrieb, V.: Improving motor rehabilitation process through a natural interaction based system using Kinect sensor. In: Proceedings of the 2012 IEEE Symposium on 3D User Interfaces, pp. 145–146 (2012)
13. VirtualRehab (2013), http://virtualrehab.info
14. Teki (2013), http://www.osakidetza.euskadi.net
15. Toyra (2013), http://www.toyra.org
16. Brontes Processing (2013), http://www.brontesprocessing.com
17. Reflexion (2013), http://www.westhealth.org/institute/our-priorities/reflexion
18. KineLabs (2013), http://www.polyu.edu.hk/kinelabs

Lateral Fall Detection via Events in Linear Prediction Residual of Acceleration

F.H. Aysha Beevi, C.F. Pedersen, S. Wagner, and S. Hallerstede

Aarhus University, Department of Engineering, Section of ECE, Denmark
{feay,cfp,sw,sha}@eng.au.dk

Abstract. Lateral fall is a major cause of hip fractures in elderly people. An automatic fall detection algorithm can reduce the time to get medical help. In this paper, we propose a fall detection algorithm that detects lateral falls by identifying the events in the Linear Prediction (LP) residual of the acceleration experienced by the the body during a fall. The acceleration is measured by a triaxial accelerometer. The accelerometer is attached to an elastic band and is worn around the test subject's waist. The LP residual is filtered using a Savitzky-Golay filter and the maximum peaks are identified as falls. The results indicate that the lateral falls can be detected using our algorithm with a sensitivity of 84% when falling from standing and 90% when falling from walking.

1 Introduction

A fall is an event which results in a person coming to rest inadvertently on the ground or floor. The fall may be caused by a disruption of a normal or expected walking gait that results in a loss of balance or it may be caused by a sudden loss of consiousness. Falls are the second leading cause of accidental or unintentional injury deaths worldwide [1].

Age plays a major role in the severity of injury due to fall. The risk rate increases with age. Older people have the highest risk of death or serious injury due to incurring a fall. A hip fracture can leave the eldely immobile. Elderly are more susceptible to lateral fall due to the impaired ability to control the postural balance in the lateral plane [2]. The risk of casualty increases if there is a delay in getting medical help. This delay can be minimized using an intelligent fall detector. The fall detector can be integrated with an Ambient Assisted Living (AAL) ecosystem. Medical sensors and actuators support AAL systems to offer personalized healthcare and wellness services [3].

Falls can be classified as forward fall, backward fall, lateral fall to the right, lateral fall to the left, syncope and neutral [4]. Yu et al. classifies falls as falls from sleeping, falls from sitting, falls from walking or the standing on the floor and falls from supports [5]. Mubashir et al. classify different types of falls as falls from walking or standing, falls from standing on supports, falls from sleeping or lying in the bed and falls from sitting on the chair [6]. Noury et al. divide the fall activity into a prefall phase, a critical phase, a postfall phase and a recovery phase [7]. The prefall phase is where a person performs normal activities. In

C. Ramos et al. (eds.), *Ambient Intelligence - Software and Applications*,
Advances in Intelligent Systems and Computing 291,
DOI: 10.1007/978-3-319-07596-9_22,

the critical phase the body moves suddenly towards the ground, ending with an impact on the ground. In the post fall phase the person remains lying on the ground and during the recovery phase the person recovers from the fall by standing up on his own or with the help of someone else. The fall is critical if the post fall phase is too long. In this work we focused on lateral falls from the walking or the standing activity.

In this paper we present a lateral fall detection based on the events in the linear prediction (LP) residual of acceleration. Our method focuses on the lateral fall detection using the acceleration in the critical phase and the postfall phase of the fall. The acceleration of the body during the fall is measured using a waist worn accelerometer. The waist was chosen as the location for the accelerometer due to the following three reasons. First, the impact forces resulting from lateral fall majorly affect the shoulder and the hip [8]. Second, Noury et al. found that the waist is the most preferred location among elderly for fall detection systems [9]. Third, the accelerometer gives the most useful measurement of the body movement when worn around the waist [10]. The acceleration measured by the accelerometer is composed of acceleration due to gravity and the acceleration due to body movement. There are many studies that focus on fall detection using waist worn triaxial accelerometer [11]. Acceleration based fall detection systems use threshold based methods or machine learning methods to detect falls [12]. Nathasitsophon et al. use information based on an LP model of the acceleration from an accelerometer to classify the activity signal as a fall [13]. Our approach is different as we use the events in filtered LP residual to detect the falls.

There are many fall detection algorithms that are, however not accurate when tested on real falls. A study by Bourke et al. evaluates the accelerometer based fall detection algorithms on real falls [14]. The study found that 13 of the published algorithms had lesser sensitivity and specificity than reported by the authors when they were tested with real fall data. These algorithms also generate more false alarms, meaning incorrectly detected fall events, when they are tested on real fall datasets. Hence there is a need for fall detection algorithms which has high sensitivity and specificity with regard to real falls.

Lateral fall is a movement of the trunk in the lateral plane from upright or sitting to a reclined or lengthened position due to a sudden acceleration experienced by the body. A common cause of fall among elderly is ability impairment to control postural balance stability in the lateral plane of motion. An impaired lateral stability causes fall that involve lateral body motion. Lateral impact has a relative risk of the order of magnitude of 2.6 compared to frontal impact for traumatic brain injury [15]. Studies have shown that right greater trochanter hit directly on the floor with a large impact in the lateral falls [16]. The acceleration experienced by the body is directly proportional to the impact force. During the critical phase of the lateral fall the inertial acceleration due to the fall will be equal to the gravitational acceleration. The direction of vector for acceleration due to gravity is upwards and the direction of inertial acceleration vector is downwards. At the end of the critical phase the acceleration increases sharply due to the impact caused by hitting on the floor.

In this study we have considered lateral fall towards right and left direction from standing height with hip flexion and knee flexion. In the types falls that we focus the falling person performs standing or walking activity in the prefall phase. At the end of the critical phase of the fall the lateral aspects of the body parts hit the ground. The body parts that impact the floor are head, shoulders, arm, elbows, wrists, thorax, pelvis, greater trochanters, thighs, knees and heels. The fall ends with the falling person lying on the floor on left side if the fall was the lateral fall towards left and right side if the fall was the lateral fall towards right.

2 Proposed Method

Proposed time-domain fall detection method

1. Compute the linear prediction residual for a 10-order forward predictor.
   - LP residual detrend and emphasize abrupt events in the signal.
2. Apply Savitzky-Golay filter to the linear prediction residual.
   - The filtering increases the signal-to-noise ratio.
3. Identify the peaks in filtered residual using nearest neighbour comparison.
   - Abrupt changes in acceleration may indicate a fall event.
4. Apply rectangualar windowing and detect falls.
   - In each window find the difference between the maximum and minimum residual peaks.
   - Determine the threshold based on the standard deviation of the window.
   - If the difference is greater than the threshold, the event is classified as a fall.
   - Move the window forward without overlapping and a step size of 51.

In our method we use Linear Prediction to predict the acceleration signal from the past 10 samples. The linear predictor model predicted the acceleration x(t) at time t using past p samples [x(t-1), x[t-2], .., x[t-p]) as

$$\hat{x}(t) = \sum_{k=1}^{p} a_k x(t-k)$$

where a_k is the prediction coefficient and t is the discrete time index. The linear prediction residual is the difference between the actual sample value and the predicted sample value.

$$e(t) = x(t) - \hat{x}(t)$$

The linear prediction residual was smoothened using Savitzky-Golay filter. We divided linear prediction residual data in time slices of length 1 second and the fall classification was done as mentioned in step 4 of the method. In the critical phase of the fall there is a short dip in the acceleration due to the free fall. It is followed by a sharp increase in the acceleration due to the body impact on the floor. In the postfall phase there is a dampening effect. This results in a large difference between the predicted acceleration and the actual acceleration. The maximum residual peak represents an actual acceleration greater than the predicted acceleration and a minimum residual peak represents a predicted acceleration less than the actual acceleration.

3 Experiment

The proposed method was tested using a data collected from a waist worn Inertial Measurement Unit (IMU) (Model : Shimmer3). The accelerometer on this wireless sensor platform is a triaxial accelerometer. The accelerometer was calibrated using a standard calibration procedure. The acceleration data was acquired at a sampling rate of 51.2 Hz and was transmitted via Bluetooth to a PC for further processing. All the signal processing and data analysis were done using Matlab.

Eleven healthy subjects in the age group of 22 to 49 weighing 63 to 110 Kg participated in the acquisition of the simulated data set. Eight participants were Judo students trained to fall. The subjects were demonstrated how to fall. The subjects did a self initiated fall from standing height to a 5 cm thick mat. The fall ended with the subject lying on the mat on their side with legs straight. The data collection was done indoor. The test subjects performed lateral fall right or lateral fall left after either walking or standing. Each fall activity of the subject's preference was repeated 5 times. The data set was made for lateral falls ending with lying flat and lateral falls ending with recovery. The IMU was worn around the waist of the subjects using an elastic belt.

4 Results

The result is true positive (TP) if the fall event is correctly detected, false negative (FN) if the fall event is not correctly detected, false positive (FP) if a non fall activity is detected as a fall and true negative (TN) if a non fall activity is not detected as a fall .

Table 1 represents the number of falls tested, the number of subjects participated, and the number of TP, FP, FN and TN as output of our method. TN is not applicable as the number of non fall activities not detected as fall are difficult to quantify. Fig.1 represents the result of a walking activity followed by a lateral fall. The figure shows the normalised acceleration of a walk activity followed by a fall. Besides acceleration it shows the filtered residual of the acceleration. The highest peak in the residual signal represents the fall activity.

Table 1. Result of Lateral Fall Classification using LP Residual

Fall Types	# Falls	#Subjects	TP	FP	FN	TN
Stand and Lateral Fall	45	9	37	7	8	N/A
Walk and LateralFall	10	2	9	1	1	N/A

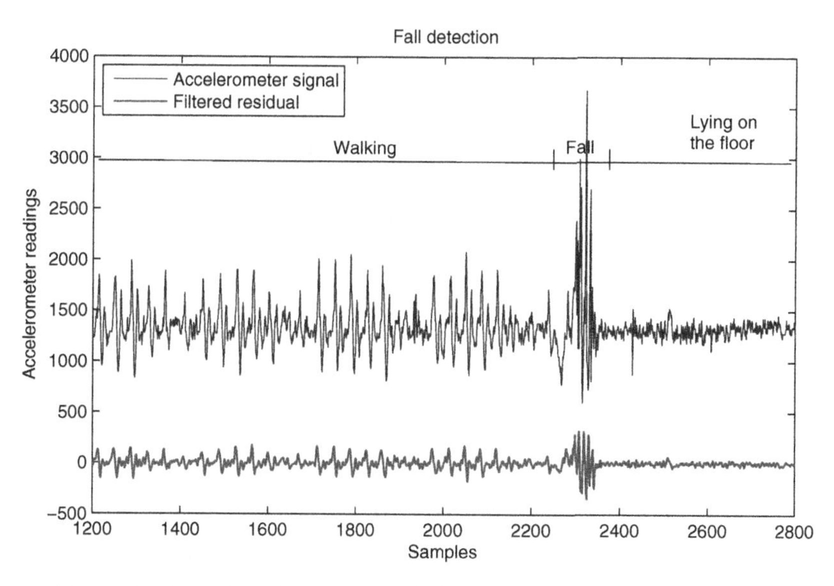

Fig. 1. Normalised acceleration signal and the linear prediction residual of a test subject walking followed by a lateral fall and the test-subject eventually lying on the floor

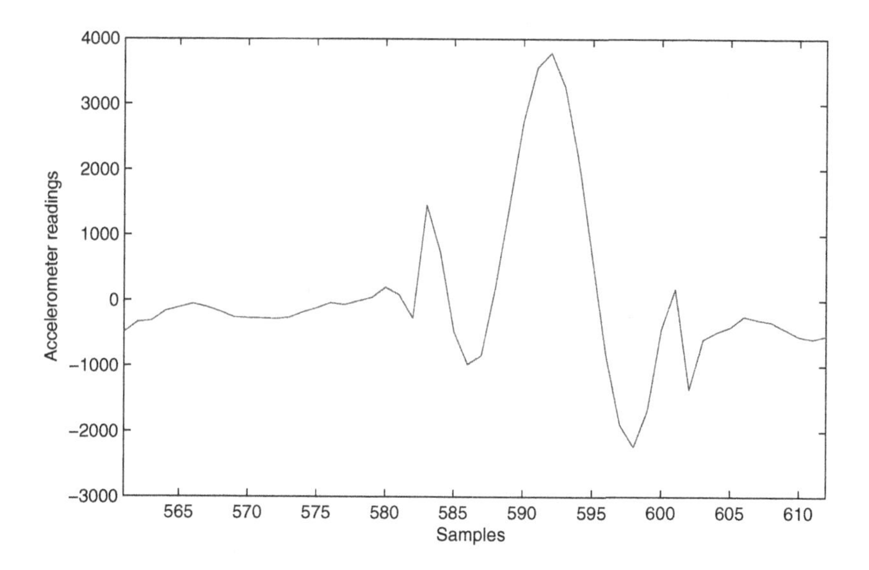

Fig. 2. Linear Prediction Residual of a fall

Fig.2 depicts the linear prediction residual in a rectangular window of size 51 during the critical phase of the fall. Fig.3 represents the acceleration and the inear prediction residual resulted from a standing followed by a lateral fall without recovery. Fig.4 illustrates the acceleration and linear prediction residual resulted from a standing followed by a lateral fall with quick recovery.

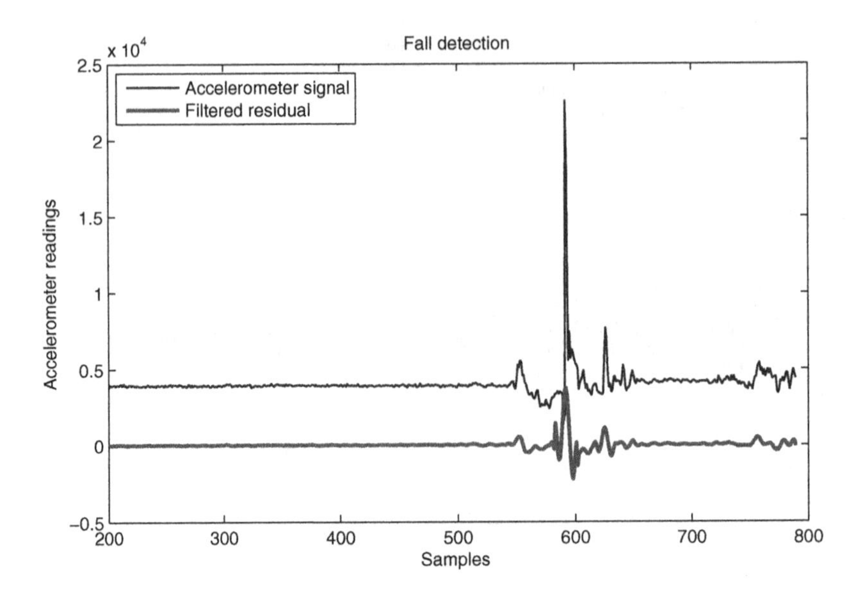

Fig. 3. Normalised acceleration signal and the linear prediction residual of a test subject standing followed by a lateral fall and the test subject eventually lying on the floor without recovery

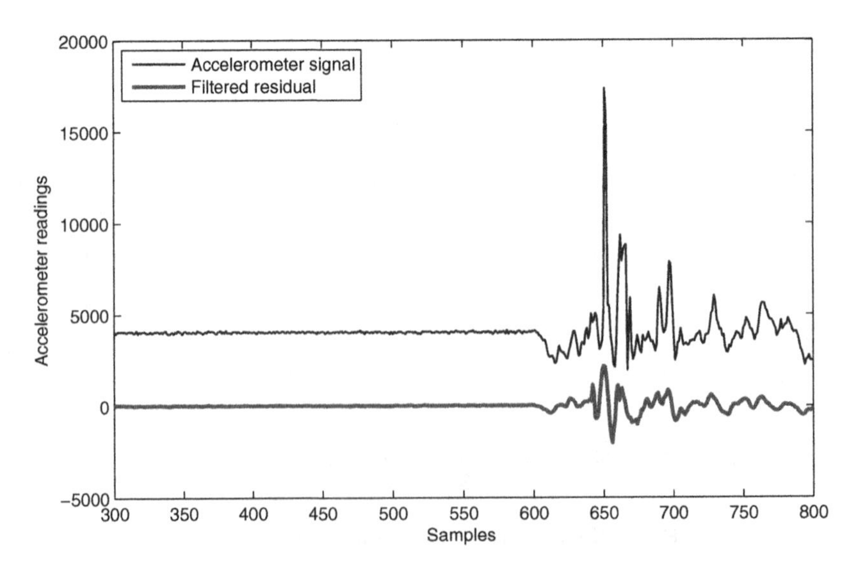

Fig. 4. Normalised acceleration signal and the linear prediction residual of a test subject standing followed by a lateral fall and the test subject eventually lying on the floor with quick recovery to a standing posture

5 Discussion

The results of the binary classification method we proposed indicate that the residual signal of LP can be used as the main source to indicate the fall event. As the acceleration series of fall activity are characterized by sharp discontinuity it causes large errors in the computed LP residual. Many of the existing threshold based fall detection methods use the acceleration sample only at the instance of the fall. But in our work we use the acceleration series of 10 samples before the instance of the fall to identify the event. Also the LP residual is not limited by the local characteristics of the signal. Hence it may be argued that probability models are more suitable for fall detection as compared to threshold based methods. Many fall detection algorithms including the LP based algorithm by Nathasitsophon et al. identify walk activity as a fall activity. In fact our method did not identify any tested walk activity as a fall activity. This is the marked difference between our methods and the method propsed by Nathasitsophon et al. Also, in their study they used a prediction order of 100 and we used a prediction order of 10 and hence our method consumes less resources. In the future the method proposed will be further investigated to reduce the FP and FN. There is no publicly available real fall dataset. As next step we will generate a model for generating a synthetic data set that closely resembles the real fall data set. The experiment presented in this paper is done as a preliminary step to develop an accurate and portable fall detector for the CareStore AAL platform.

Acknowledgement. This study is funded by the CareStore project (www.carestore.eu) funded by the European Commission under the gant agreement no. 315158.

References

1. Ageing, W.H.O., Unit, L.C.: WHO global report on falls prevention in older age. World Health Organization (2008)
2. Rogers, M.W., Mille, M.-L.: Lateral stability and falls in older people. Exercise and sport sciences reviews 31(4), 182–187 (2003)
3. Memon, M., Wagner, S., Hansen, F.O., Pedersen, C.F., Aysha, F.H., Mathissen, M., Nielsen, C., Langvad, O.: Ambient assisted living ecosystems of personal healthcare systems, applications, and devices. In: Scandinavian Conference on Health Informatics 2013, p. 61 (2013)
4. Noury, N., Fleury, A., Rumeau, P., Bourke, A.K., Laighin, G.O., Rialle, V., Lundy, J.E.: Fall detection-principles and methods. In: 29th Annual International Conference of the IEEE Engineering in Medicine and Biology Society, EMBS 2007, pp. 1663–1666. IEEE (2007)
5. Yu, X.: Approaches and principles of fall detection for elderly and patient. In: 10th International Conference on e-health Networking, Applications and Services, HealthCom 2008, pp. 42–47. IEEE (2008)
6. Mubashir, M., Shao, L., Seed, L.: A survey on fall detection: Principles and approaches. Neurocomputing 100, 144–152 (2013)

7. Noury, N., Rumeau, P., Bourke, A., ÓLaighin, G., Lundy, J.: A proposal for the classification and evaluation of fall detectors. Irbm 29(6), 340–349 (2008)
8. Lo, J., Ashton-Miller, J.: Effect of pre-impact movement strategies on the impact forces resulting from a lateral fall. Journal of biomechanics 41(9), 1969–1977 (2008)
9. Noury, N., Galay, A., Pasquier, J., Ballussaud, M.: Preliminary investigation into the use of autonomous fall detectors. In: 30th Annual International Conference of the IEEE Engineering in Medicine and Biology Society, EMBS 2008, pp. 2828–2831. IEEE (2008)
10. Mathie, M., Celler, B.G., Lovell, N.H., Coster, A.: Classification of basic daily movements using a triaxial accelerometer. Medical and Biological Engineering and Computing 42(5), 679–687 (2004)
11. Bourke, A., Van de Ven, P., Gamble, M., OConnor, R., Murphy, K., Bogan, E., McQuade, E., Finucane, P., OLaighin, G., Nelson, J.: Evaluation of waist-mounted tri-axial accelerometer based fall-detection algorithms during scripted and continuous unscripted activities. Journal of biomechanics 43(15), 3051–3057 (2010)
12. Igual, R., Medrano, C., Plaza, I.: Challenges, issues and trends in fall detection systems. Biomedical Engineering Online 12(1), 66 (2013)
13. Nathasitsophon, Y., Auephanwiriyakul, S., Theera-Umpon, N.: Fall detection algorithm using linear prediction model. In: 2013 IEEE International Symposium on Industrial Electronics (ISIE), pp. 1–6. IEEE (2013)
14. Bagalà, F., Becker, C., Cappello, A., Chiari, L., Aminian, K., Hausdorff, J.M., Zijlstra, W., Klenk, J.: Evaluation of accelerometer-based fall detection algorithms on real-world falls. PloS one 77(5), e37062 (2012)
15. Vavalle, N.A., Moreno, D.P., Rhyne, A.C., Stitzel, J.D., Gayzik, F.S.: Lateral impact validation of a geometrically accurate full body finite element model for blunt injury prediction. Annals of Biomedical Engineering, 1–16 (2013)
16. Nankaku, M., Kanzaki, H., Tsuboyama, T., Nakamura, T.: Evaluation of hip fracture risk in relation to fall direction. Osteoporosis International 16(11), 1315–1320 (2005)

Fighting Elders' Social and Technological Exclusion: The TV Based Approach

Luís Correia[1,2], Nuno Costa[1,2], and António Pereira[1,2]

[1] School of Technology and Management, Computer Science and Communication Research Centre, Polytechnic Institute of Leiria, Leiria, Portugal
[2] Information and Communications Technologies Unit, INOV INESC Innovation-Delegation Office at Leiria, Leiria, Portugal
luisfjcorreia@hotmail.com,
{nuno.costa,apereira}@ipleiria.pt

Abstract. We are assisting to the fastest grow of senior population ever and that tendency has brought several challenges for governments, families and for the elderly. The society was not prepared for that. On the one hand, there are many elderly who live alone in the cities and, on the other hand, active people is moving to the cities looking for a better life while leaving behind the villages where they born and grow and the older family members. On the other side, everyone is expecting the aid of technology in order to solve or at least minimize this problem. This paper present a video-calling service targeted for elderly social and technological exclusion and promoting socialization while using recent technology embedded into well-known electronic devices like TVs. The evaluation of results showed that when assisted technology is encapsulated into everyday objects, older people can use it seamlessly, without any learning curve.

Keywords: HbbTV, Video call service, Elders.

1 Introduction

Nowadays, we are assisting to the fastest growing of older population. This trend is even more remarkable at rural areas where in some cases 100% of the population is elderly. This tendency comes from several different factors like born ratios, increasing aging, active people moving to the cities seeking a better life while leaving behind older family members. This situation leads to few population residing in rural areas, whereas in its most elderly, suffering from social exclusion, often ending up living alone [1]. This exclusion is not only social, but also technological, because, on the one hand, older people have no suitable conditions to learn about technology, and, on the other hand, many times the village has no technological infrastructures like cable TV, Internet broadband access or even cellular network as the number of people is not economically profitable. This technological disinterest contributes to the increase of technological ignorance, now present in much of the elderly population [2].

C. Ramos et al. (eds.), *Ambient Intelligence - Software and Applications*,
Advances in Intelligent Systems and Computing 291,
DOI: 10.1007/978-3-319-07596-9_23,

Media and research community argue that technology can solve this situation or at least minimize it by supplying means to approximate people. However, older people don't have knowledge about interacting with technology and most of the times in their villages there is no support for learning that. It was this assumption that inspired us to develop technology to fight social and technological exclusion using technology embedded into electronic everyday objects without requiring the use of a traditional computer. As the most used electronic device at elders home is the television, it was decided to embed the solution on the TV as elderly already knows how to interact with it, without requiring any learning curve.

This paper describes a video-call service through TV, which facilitates the conversation with family members, friends and service professional like doctors, plumbers, electricians, among others. The service offers an extremely simple and intuitive interface where technological knowledge is not required at all.

The paper is organized as follows: section 2 describes the related work, section 3 presents and details the system architecture while section 4 details the system implementation. Section 5 describes the usability evaluation tests and results and the paper is concluded in section 6.

2 Related Work

This section surveys technological solutions and projects in the scope of social and technological exclusion of the senior population.

TV-kiosk [3] is a project that aims to stimulate interaction among senior users through technology in order to minimize isolation. This project relies in a decentralized architecture in order to avoid the single point of failure and in a Virtual Private Ad Hoc Network to secure communications. The interaction of users with the platform is carried out through television, which offers a simple interface and a remote control. Although in the scope of fighting social and technological exclusion experienced by senior population, this solution does not offer video calling.

The work present in [4] was carried out in order to evaluate the most efficient approaches to issue video calls through IPTV services by using a IP Multimedia Subsystem (IMS) base architecture. To achieve that, authors relied into two approaches of coding and transmission video calls: i) using a set-top-box and ii) using a digital video camera.

Authors concluded that set-top-box base coding is slower than camera coding presenting a delay of about ¾ of a second. However, the lower price of a ready for coding camera is about 150 dollars. This work is very important for this paper, because it compares the different approaches to transport and encoding video stream and analyzes the advantages and disadvantages of each.

In [5], authors seek enrichment of IPTV service, by integrating new communication services, and presented two main use cases: i) shared user experience and ii) cross-service e-commerce. In the first case, the authors integrated a multimedia communication approach relying in *Voice over Internet Protocol* (VoIP) to aid users sharing reactions and emotions among themselves while watching TV programs.

The authors argue that videoconference and voice email are also good services to be adopted. In the second case, authors proposed an e-commerce platform using TV to access it. To do that, authors argue that TV operators and commercial organizations must be in partnership. In practice, the user would receive publicity in the TV screen and a buy option using a reduced number of steps.

Evolving IPTV Service architecture [6] is the Cisco Service Exchange Framework that allows IMS and non IMS services integrate with IPTV, in order to support call reception through TV or programmed recording sessions using a smart phone.

The work described in this paper is similar to the ones surveyed in this section but has a distinct capability, the issue or reception of video calls while offering an extremely simple and intuitive interface as it is based on a TV remote control like device.

3 Architecture

As already referred, the solution described in this paper uses the TV as output device and the TV remote control as the input device. Once new buttons are required it was decided to mimic TV remote control using a smart phone. In order to integrate TV service and other services offered through the Internet, it was decided to adopt the HbbTV standard. Figure 1 shows a high level overview of the proposed system architecture. The main modules of this architecture will be presented in the next sections.

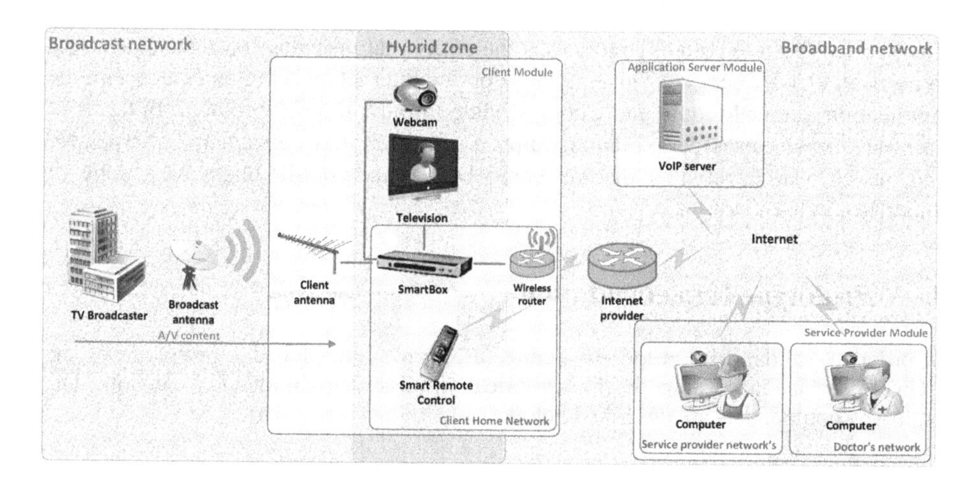

Fig. 1. General architecture

3.1 VCoTV Service

The Video Call on TV (VCoTV) service aims to help elderly issuing video calls using TV and TV remote control. This service was developed having in mind the limitations of elderly. However, this service could be used by anyone. In terms of functionalities, the service deals with TV contents and video calls. As far as TV is concerned, VCoTV service must support channel changing, volume control and contents guide. By its turn, the video call functionality must support the issuing and receiving of video calls, show contacts and contact listing management. It must be possible to deal with video calls while watching TV without any interruption.

According to figure 1, the VCoTV service it's divided into client module and application server module. In the client module, the VCoTV runs in the SmartBox where a web camera and a microphone are also connected. In terms of software, the client side of the VCoTV service was implemented using Session Initiation Protocol (SIP), which is connected to the Applicational Server Module. This allows client side to issue calls through the server side. Additionally, all service providers available through video call (see service provider module) are also connected to Application Server Module in order to be reached by elderly.

3.2 The Smart Remote Control

Considering the requirements of service and the limitations of current remote controls, it was decided that Smart Remote Control is an Android application. Beyond sending commands to TV, like a TV standard remote control, this provides buttons to manage video calls. The Smart Remote Control interacts with SmartBox through WiFi by using TCP sockets.

3.3 Service Provider Module

The Service Provider Module represents the entity that provides commercial services to users of VCoTV service. This module is on the Internet and communicates with the Application Server Module and Client Module through the SIP protocol. In terms of hardware, service provider module requires a computer with webcam and microphone and, at software level, this module could be a standard SIP client or a copy of SmartBox VCoTV software.

4 Prototype Implementation

In order to validate the proposed architecture, it was developed a prototype of the VCoTV service according to the specification described in the last section. The prototype components are described below.

4.1 Application Server Module

The Application Server Module supplies the VoIP service to both VCoTV module and Service Providers Module. The Application Server Module is based in Trixbox CE (version 2.8.0.4), which allows installation, configuration and management of Asterisk Private Branch Exchange (PBX) VoIP solution. In the developed prototype, the Application Server Module runs in a laptop.

4.2 Client Module

Client Module includes both the VCoTV which runs in the SmartBox and the Smart Remote Control which runs in an Android smart phone.

Client module handles TV contents and video calls by allowing changing TV channel, change volume, show TV contents guide, issue video calls, receive video calls and contact listing management.

In terms of hardware the client module relies on a Raspberry Pi, B model as it is a cheap piece of hardware which includes all necessary port interfaces. For digital signal decoding it was used a AVERMEDIA AverTV Volar HD PRO A835 device due to its dimensions, energy consumption and USB support. For image and sound capturing, it was used a standard Webcam with integrated microphone. Additionally, it was also used an indoor UHF antenna to capture Digital Terrestrial Television signal and a wireless router to link all the network components.

To implement the VCoTV service it was used Python, Qt Meta Language (QML) and Qt Framework to build the graphical interface, VLC to access DVB-TV board, Omxplayer which supports hardware acceleration to reproduce multimedia contents on TV and PJSIP for issue and reception of video calls.

During the development of the graphical interface, the following specified requirements were accomplished:

- Simple interface with minimal information;
- Usage of big text fonts (minimal size of 25px);
- Background, foreground and text color contrast according to the WCAG 2 AA (5.04).

4.3 Smart Remote Control Module

The Smart Remote Control Module is an Android Application that mimics a standard TV remote control. This application has a very simple and intuitive interface to allow elderly interact with SmartBox. During the development of the graphical interface, the *Web Content Accessibility Guidelines* 2.0 (WCAG) specification was followed.

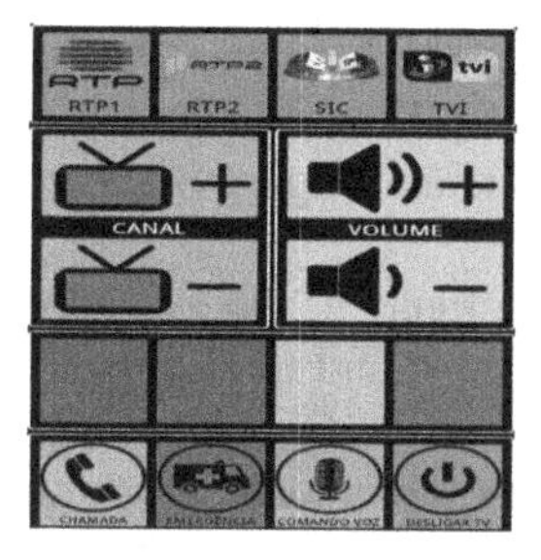

Fig. 2. Smart Remote Control interface

As can be seen in figure 2, the smart remote control interface has five groups of buttons:

- On the top, there is the four public open TV channels;
- On the second and third rows, on the left, there is the channel buttons and on the right the volume buttons;
- On the fourth row, there are the context buttons which will be used to interact with the VCoTV module. These buttons have no specific meaning as it depends on the VCoTV menu position;
- The fifth row includes the video call button, the emergency button that can be used to issue a call to a family member or hospital, the voice command button to interact with the VCoTV service using voice and the turn off TV button that also shuts down the VCoTV software in the smart box.

5 Evaluation

The prototype described in this paper, was evaluated in the performance, usability and quality of experience perspectives. Due to the limit of paper pages of this paper, in this section only the usability evaluation is described.

To evaluate the usability perspective, it was asked to a group of elderly to use/test the prototype. The authors explained the smart remote control buttons and defined a list of operations to be executed by elderly. The list of operations included the following actions:

1. Change to the next channel;
2. Decrease the volume level;
3. Change to the previous channel;
4. Change to full screen;
5. Read contents guide;
6. Answer call;
7. List contacts;
8. Issue an emergency video call.

The usability tests took place in a nursing home in a group of five persons, two males and three females with ages in the 77 to 91 interval with an age average of 82 years old, where all the users were literate. Figure 3 shows an old person using VCoTV service.

According to the graph in the figure 4, the execution time for the first operation is high (7.6 seconds), while the second and third operations took less time (7.2 and 5,4 seconds respectively). In the set of the first three operations, the third is accomplish with less average time because users memorize the button positions in the smart remote control, while the fourth operation takes the bigger time because this operation requires the user attention to the TV screen while accomplishing it.

Fig. 3. Usability evaluation by elders

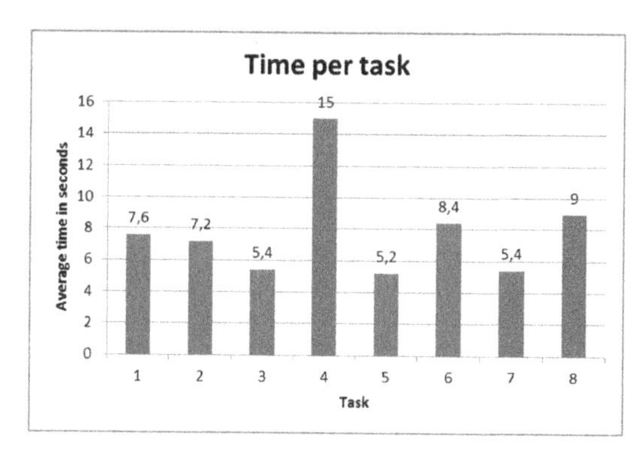

Fig. 4. Average time for each operation

When accomplishing fifth and seventh operations, users are already familiar with the TV screen interface and finished the operations quickly. Sixth and eight operations have a high accomplish time because users have several difficulties understanding which button to push in order to finish the operation.

6 Conclusion

This paper describes the work carried out in order to fight both social and technological exclusion mainly focused on elderly. The proposed solution gives elderly the opportunity to use ultimate Internet services encapsulated into everyday electronic object, the TV. This contributed to eliminate barriers between elderly and technology, lower the learning curve and promotes the development of new services based on the proposed architecture.

The usability evaluation of the developed prototype showed interesting results as elderly with age average of 82 years old, even those without technological knowledge, can use the prototype without any difficulty. Despite that, one limitation was identified in the performance evaluation tests (not described in this paper) due to space constraints. Performance tests have showed that a raspberry Pi based SmartBox does not offer the required response time for full screen and channel changing operations. These two operations took about 15 seconds and must be optimized in the future.

References

[1] Instituto Nacional de Estatística, Censos 2011 – Resultados Provisórios, Edição 201 (2011)
[2] Alm, N., Gregor, P., Newell, A.F.: Older people and information technology are ideal partners. Dundee, Scotland (2002)
[3] Steenhuyse, M., Hoebeke, J., Ackaert, A., Moerman, I., Demeester, P.: TV-kiosk: An Open and Extensible Platform for the Wellbeing of an Ageing Population, pp. 54–63. Ghent, Belgium (2012)
[4] Zeng, H., Peng, X.-B.: Research and Implementation of IPTV video calls based on the IMS network. In: International Conference on Computer Science and Service System, Nanjing, China, pp. 1059–1062 (2012)
[5] Neuwirt, O., Da Silva, J., Abbadessa, D., Winkler, F.: Towards a New User Experience in IPTV: Convergence Services and Simpler E-commerce on IMS-based IPTV (2008)
[6] Cisco Public Information, The Evolving IPTV Service Architecture (2007)

A Decision Support System for Medical Mobile Devices Based on Clinical Guidelines for Tuberculosis

Sílvio César Cazella[1,2], Rafael Feyh[2], and Ângela Jornada Ben[1]

[1] Federal University of Health Sciences of Porto Alegre, Rio Grande do Sul
Rua Sarmento Leite 245, 90050-170, Porto Alegre, RS, Brazil
[2] University of Vale do Rio dos Sinos, Centro de Ciências Exatas e Tecnológicas
Av. Unisinos 950, 93022-000, São Leopoldo, RS, Brazil
{silvioc,angelajb}@ufcspa.edu.br, r_pohren@hotmail.com

Abstract. The decision making process conducted by health professionals is strongly linked to the consultations of clinical guidelines, generally available in large text files, making the access to the information very laborious and time consuming. The health area is very fertile for the emergence of solutions based on mobility. Among others, there are solutions that offer to the doctor the access to the right information when and where needed, contributing so that the health services can be performed at anytime and anywhere. This study presents an application for mobile devices where you can get access to the content of clinical guidelines (in this case Tuberculosis' clinical guideline). The prototype was evaluated by medical experts, showing promising results.

Keywords: Mobile Devices, Clinical Guidelines, Android.

1 Introduction

The use of mobile devices is increasingly present in the lives of professionals in various fields. In particular the area of health, because it is a very sensitive area due to their needs and relevance. The application of mobile technology in healthcare will enable their professionals to be more productive and efficient, bringing significant benefits to the staff of Health and their patients.

Mobile applications can connecting patients and providers related to health services, making the patient's history easier monitoring, once the patient has medical follow in his hands. Thus, mobile applications will play an important role in shaping the future of systems of public and private health. The steady increase in the use of mobile devices such as mobile phones, smartphones, tablets, among others, by health professionals has minimized the resistance of the users regarding the use of such applications. Thus, increasingly we find applications available for these devices, which in turn become excellent tools for the working environment of health professionals. In 2013, for the first time in history , smartphone sales surpassed sales of regular mobile phones , reaching , based on data from the first quarter , the mark of 216 million units , representing 51.6 % of total mobile phones and smartphones marketed during this period [7].

C. Ramos et al. (eds.), *Ambient Intelligence - Software and Applications*,
Advances in Intelligent Systems and Computing 291,
DOI: 10.1007/978-3-319-07596-9_24,

According to [3] , the total number of smartphones sold worldwide in the second quarter of 2013 was 236.4 million units , representing an increase of 51.3 % compared to the second quarter of 2012 . The Android operating system maintained its leadership position , achieving a growth of 73.5 % in the second quarter 2013 compared to the same period of 2012 and was present in 79.3 % (market share) of smartphones marketed [3].

The health area is a fertile ground for the emergence of solutions based on mobility, once these solutions remove the geographical, temporal and physical barriers, providing the physician access to information where and when needed, allowing health services can be delivered anytime and anywhere [5].

It is noticed that the use of mobile devices plays a very important role in supporting health services , providing greater flexibility from simple processes such as data collection , up to the use of systems to assist in the decision making process with different levels of complexity [2]. In the health one of the global challenges is combating tuberculosis. Only in Brazil is estimated that 4500 people die each year, victims of this disease, which is considered curable and preventable. Even with today's technological advances that are already able to control it, it will be necessary to develop new vaccines and drugs for the elimination of this disease [1].

Taking into consideration all the points described and the motivation of this research , the objective of this work was: - propose an application for mobile devices to facilitate the process of medical decision making, in this case study was used a clinical guideline for tuberculosis in Brazil. The remainder of the paper is organized as follows: The second section contains the theoretical content used in building the knowledge required for the development of this work . In section three the methodology is described. The prototype is detailed in section four. In section five the evaluation performed and results are discussed.

2 Mobile Health and Tuberculosis' Guideline

Worldwide, providing services in healthcare has been transformed by the use of mobile and wireless technologies. A powerful combination of factors is driving this change, where the rapid advances in technologies and mobile applications favored by continued growth in coverage of mobile cellular networks, have increased the number of opportunities for the integration of mobile health services with existing health today [8].

The unprecedented spread of mobile technologies as well as advances in its use to address the challenges of Health contributed to the emergence of mobile health or mHealth as it is known internationally. In the next subsections we discuss a bit more about mHealth and Tuberculosis (disease used as case in this research) and the clinical guideline.

2.1 mHealth

The Health Industry is moving towards an architecture of distributed services, enabling important decisions happen directly at the point of care, the main engine of change , mobile Health (mHealth) [5].

According to the Global Observatory for eHealth(GOE) , the meaning of mHealth is the use of cell phones, smartphones, tablets , personal digital assistants or any other mobile device , such as support to public policies and activities related the area of health [8].

Currently the global mHealth market presents strong growth, overcoming countless barriers created by the global crisis , and may reach the figure of 26 billion dollars in 2017 . This figure makes clear that the mHealth conquered its space on the world stage assistance and medical care , steadily become a tool for transformation of health systems , currently there are almost 97,000 mHealth' applications available worldwide [4]. One of the major problems faced today by health professionals is access to quality information where and when needed. To address this difficulty , the Health industry in mHealth found a powerful ally , able to provide the necessary access people wherever they are [5].

Some platforms have become benchmarks for the development of applications for mobile devices , and among these is the Android Platform [6] that was used in prototyping the application of this work.

2.2 Tuberculosis

In 2003 , tuberculosis was elected by the Ministry of Health in Brazil as a priority public health problem being tackled. That same year, the budget of the National Tuberculosis Control Programme (NTCP) was increased by more than 14 times , in addition to technical and administrative measures such as encouraging the participation of civil society in controlling this disease and the expansion and upgrading of the NTCP team [1].

As a guideline for tuberculosis in Brazil , there is the " Manual of Guidelines for Tuberculosis Control in Brazil ," published in 2011 by the Brazilian Ministry of Health . This document consists of a manual of recommendations that should guide medical management related to tuberculosis , to be followed by health professionals within the Unified Health System [1].

3 Methodology

It is an applied research, which is to use the acquired knowledge to solve real problems of the current context. As for his approach was qualitative and quantitative. The research methods that guided this work were bibliographical research, case study and prototyping. This research involved the participation of an expert in the area of health, who assisted in matters relating to this area. The case study was based on the Brazilian guidelines for tuberculosis, more specifically, the "Manual of recommendations for the control of tuberculosis in Brazil ", published in 2011 by the Ministry of Health [1].

The evaluation of the prototype was performed by three specialists, experts in primary care and in tuberculosis, through a questionnaire. Thus, it was possible to identify their perception about the proposed solution. For data analysis, a quantitative and qualitative approach were applied.

4 Tuber's Prototype

When a doctor receives a patient with symptoms of tuberculosis in health care in Brazil (primary care), he needs the help of clinical guidelines that the ministry of health issued to aid your decision making. These guidelines are printed. This consultation procedure of how experienced the doctor is time consuming, and eventually affects the productivity of the doctor. Thus it was decided to develop an prototype for mHealth called Tuber. This application was developed based on main rules taken from clinical guidelines and modeled with the help of medical expert in this disease.

The architecture of the prototype was developed as a client- server architecture, therefore, is divided into two packages : 1) as a client, we have an application for mobile devices with Android operating system (hereafter in the text this article will be referred to as mobile application), and 2) the server side, which is formed by a java application containing a web service (WS) that communicates with a database and provides services to the mobile application (hereinafter forward the text of this article will be referred to as backend).

The backend has the principal responsibly, providing services that enable the mobile application to access the existing database server information. For its development we used the Java and Axis2 1.6.2 programming framework. Thus, it was possible to easily build WSs in standard Simple Object Access Protocol (SOAP) , which is used protocol for exchanging structured to be moved from one device to another through WS information . As a framework for object - relational mapping opted for the Hibernate 4.2.6, by this encapsulate all interaction with the database . This Java application runs on an Apache Tomcat 7 server , and uses a MySQL Community Server 5.6 database. The mobile application was developed based on the Model View Controller (MVC), which separates the application into three layers . Therefore, following this pattern are: (a) the layout XML files representing the View layer, (b) the activities represented Controller, and (c) the classes of the model represented Model.

Even using the development platform Android 4.3 SDK, the application is designed to run on mobile devices running Android OS 2.2 or higher. As a database for the application, the option chosen was the SQLite, because this is now natively supported by Android . Until the version used in this work, Android has not had native support for interaction with WS. Thus, for the construction of the classes necessary for communication with the WS provided by the backend, the library ksoap2 3.0.0 was used.

4.1 The Clinical Guideline Tuberculosis : Case Study

To evaluate the operation of the prototype developed for this work , it was necessary create the content related with Health. Therefore,the Brazilian Guidelines for Tuberculosisas was used as case study.

The process of developing the Tuber application was strongly based in the support provided by clinical specialist, who through weekly meetings, helped directly in the definition of what parts of the guideline should be used and how they should be organized in the application. The definition of the content was focused on Primary Care to Health.

As a source of information , we chose to use, as faithful as possible , the content of the guideline. This constructive process gave rise , in addition to the description of the guideline itself through 146 items (small parts of the guideline) , divided between the two patient defined profiles: 1) " *Children under the age of ten years*" and 2) "*Adults or Teens with greater than or equal to ten years old* ". During the prototype's specification phase, the specialist cited as a difficulty encountered the need to manually perform the calculation of the scoring system used for the diagnosis of tuberculosis in children. This system consists of five variables : 1) **Clinical Situation**: if the Clinical Situation is symptomatic then is assigned 15 points, but if it is asymptomatic is assigned 0 points; 2) **Radiological Situation**: if the Radiography presents condensation and infiltration for more than two weeks then is assigned 15 points, if it is less than two weeks is assigned 5 points and if radiography is normal then is assigned 0 points; 3) **Contact with Tuberculous Adult**: brand yourself 10 points if had close contact over the past two years and 0 points if it was casual or negative ; 4) **Tuberculin Test**: if the result was greater than or equal to 5mm, is assigned 15 points, if less than 5mm is assigned 0 points , and 5) **Nutritional Status**: if the patient presenting severe malnutrition is assigned 5 points, with standard is assigned 0 points . By the sum of points obtained in each variable , we obtain the final result, where : 1) if the result is greater than 40 points the patient should starts treatment ; 2) if the result is between 30 and 40 points the patient should starts the treatment according doctor's decision, and 3) if the result is less than 30 points, the treatment is not necessary.

To help the doctor in this task, a flow was definide and this flow was based on a data structure with a tree format, where we have : 1) the questions representing nodes (Item), 2) the score obtained as edges (ITEMLIST) and 3) the final score and its interpretation as sheets (Item). Thus, to properly answer the questions (fig. 1(c)), we can navigate the created flow to reach a final result containing the diagnosis of tuberculosis in children. Similarly , other flows from the guideline were analised and new ones were also created, as an example we can mention the diagnosis performed on adults using the tuberculin skin test (TST) , the test result can be obtained by the interpretation of the information presented by a table contained in the guideline. This table is available in the mobile application through a flow of navigation to answer questions regarding the result of the TST and the patient's age, then groups of patients to be treated for Tuberculosis are displayed.

Fig. 1 shows some interfaces of the prototype, they assist the doctor during patient consultation and diagnosis regarding Tuberculosis. Figure 1(a) presents the first interface that doctor has contact when starts the Tuber's application, figure 1(b) presents the second interface where the doctor should informs if the patient is a child or an adult (this information helps the system decide what flow should be presented in the sequence), figure 1(c) presents the third interface where the doctor gives some patient informations, and has the result to take decision based on tuberculosis' clinical guideline (the guideline's page is informed on this interface).

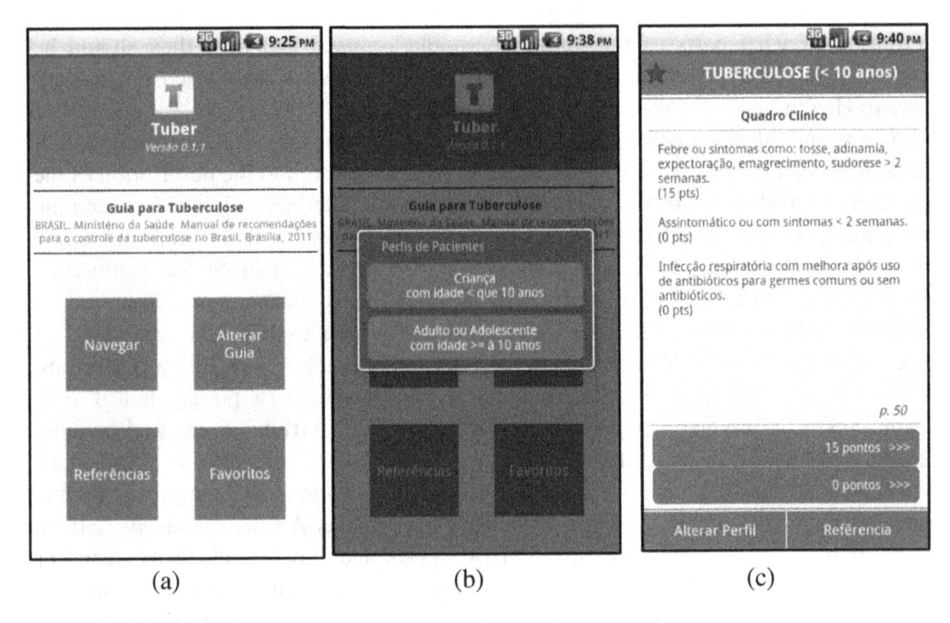

Fig. 1. Tuber's Interfaces

5 Evaluation

The evaluation of the prototype was conducted through a questionnaire. The questionnaire was applied to three medical from family and community, one being a professor of public health. The questionnaire comprises sixteen questions, which are divided into three groups.

The group formed by the questions one through six have aimed to collect data on the profile of the respondents and their familiarity with the use of applications on mobile devices . This first group consists of four questions of free choice and two free response. Having as objective the evaluation of the prototype acceptance by the respondents , we have the second group which is composed from question number seven to fourteen, comprising : (a) six questions that use a Likert scale of five points for the answer , where the value " 5 " means " Strongly Agree " and " 1 " means " Strongly Disagree " , (b) a matter of free choice , and (c) a matter of free response . Finally , in the last group we have questions 15 and 16, to collect users comments and suggestions for improvement to the application and to the " Guide to Tuberculosis ".

The sample of experts which agreed to participate in this evaluation was three. The profile of this non-probabilistic sample was composed primarily of health professionals, two of whom are also professors. All have as the practice area of family and community medicine, having between 5 and 17 years of expertise in this area.

The first group of questions investigated the familiarity of the respondents with the use of applications on mobile devices such as smartphones and tablets . With responses submitted, we concluded that all experts already were familiar with this context and already use applications targeting the area of Health on their smartphones.

The second group of questions investigated the acceptability of the developed prototype . Usability is an important factor in the acceptance of any software. Within this context were proposed questions 7, 8 and 9. All respondents agreed completely that the application is easy to understand (question 7) and easy to use (question 8) , containing clear and objective features (question 9). According to question 10 , there was a massive agreement among raters, with the possibility of the application developed replace the use of printed guideline. It was also obtained the full agreement of the respondents in question 11, where was asked about the possibility of adding and changing content directly by health professionals, thus allowing for more flexibility and consistency. A point worth mentioning because it is directly connected with the purpose of this research is the fact that all respondents totally agree (question 12) that the prototype created efficiently assists in the decision making process of the health professional.

Finishing the second group, question 13 tried to verify from the respondents if they would use this application for patient care, and all answered yes. Question 14 asked the justification for the use or not use the application.

Table 1. Contribuition by evaluators for question 14

Evaluator	Contibuition
1	*"The application has the practicality to summarize and organize the information needed and more relevant in the outpatient management of patients with suspected and diagnosed with tuberculosis. The fact of being in the cell and can be carried in the pocket also contribute to use."*
2	*"I would use the application during patient care, because it allows access to the information they need to diagnose and treat people with TB and / or latent tuberculosis infection in a clear and objective manner. This facilitates my work and optimizes patient care time."*
3	*"The application facilitates the conduct of research for each case, considering that the primary care professionals working with several other health problems and can not write everything that comes to meet that specific disease."*

6 Conclusions and Future Work

The use of mobile devices is increasing gradually , making these devices a new source for developing solutions in various technologies . In Health , mHealth is gaining more prominence by facilitating access to information when and where needed. The Tuber application confirms the feasibility of the proposed solution, and presents extensive content covering since the diagnosis of Tuberculosis up to their treatment , using information obteined from the Brazilian's Guide for Tuberculosis. All experts agreed that the application is easy to understand and use, and it assists efficiently in the process of decision making of health professionals facing the demand for treatment in health centers, allowing quick access to the information they need to diagnose and treat people with tuberculosis. The proposed solution has demonstrated to be useful for patient care, to optimize and facilitate the work process, providing more time for patient care.

After performing all the steps of this work, and especially the massive acceptance that the application obtained from your target audience, it is understood that the objective of this research has been achieved. Also mean that the solution presented here, is directly connected to an existing need for the professionals of health, who wish to streamline and facilitate the process of decision making so that they can devote more time to patient care. Completing this survey, it is possible to identify how future work : a) develop a web application that allows creation and editing guides , directly by health professionals ; b) carry out the validation of the prototype by means of a clinical trial ; c) adapt the application for use on tablets and create versions for Windows Phone and iOS platforms.

References

1. Brasil. Ministério da Saúde. Manual recommendations for tuberculosis control in Brazil. Brasília (2011), http://portal.saude.gov.br/portal/arquivos/pdf/manual_de_recomendacoes_tb.pdf
2. Grasso, M.: Clinical applications of hand held computing. In: Proceedings of the 17th IEEE Symposium on Computer-Based Medical Systems (2004)
3. IDC. Apple Cedes Market Share in Smartphone Operating System Market as Android Surges and Windows Phone Gains, According to IDC. Framingham (2013), http://www.idc.com/getdoc.jsp?containerId=prUS24257413
4. Jahns, R.-G.: The market for mHealth app services will reach $26 billion by 2017. Berlin (2013), http://www.research2guidance.com/the-market-for-mhealth-app-services-will-reach-26-billion-by-2017/
5. Krohn, R., Metcalf, D.: mHealth: from smartphones to smart systems. HIMSS, Chicago (2012), http://www.healthsen.com/mHEALTH_PREVIEW.pdf
6. Lee, W.-M.: Beginning Android application development. Wiley, Indianapolis (2011)
7. Mendiola, J.: Los smartphones superan en ventas a los móviles convencionales por primera vez. [S.I.] (2013), http://es.engadget.com/2013/04/26/smartphone-moviles-idc-q12013/
8. World Health Organizationm (WHO). mHealth: New horizons for health through mobile technologies, Geneva. Global Observatory for eHealth series, vol. 3 (2011), http://www.who.int/goe/publications/ehealth_series_vol3/en/index.html/

New Applications of Ambient Intelligence

Davide Carneiro and Paulo Novais

CCTC/DI - Universidade do Minho
Braga, Portugal
{dcarneiro,pjon}di.uminho.pt

Abstract. Ambient Intelligence emerged more than two decades ago, with the exciting promise of technologically empowered environments that would be everywhere, cater to all our needs, be constantly available, know who we are and what we like, and allow us to make explicit requests using natural means instead of the traditional mouse and keyboard. At a time in which this technological unravelling was expected to have already happened, we still use the mouse and the keyboard. In this paper we make a brief analysis of why is this evolution taking more than initially expected. We then move on to analyse several different projects that are innovative, in the sense that they encompass fields of application that go beyond the initially envisioned, and show the diverse areas that AmI systems may potentially come to change.

Keywords: Ambient Intelligence, Future, Innovative Scenarios.

1 Introduction

In 2001, [5] described four scenarios for Ambient Intelligence, that depicted "what living with 'Ambient Intelligence' might be like for ordinary people in 2010". In the first scenario, 'Maria' – Road Warrior, Maria is someone who is travelling in Europe with a very lightweight hand baggage and no "traditional" personal computing devices: all these have been replaced by the 'P-Com' on her wrist. This device identifies Maria and takes care of going through the airport security, arranges for a rented car, carries her preferences so that the hotel room can adapt to her, among other features. In Scenario 2, 'Dimitrios' and the Digital Me' (D-Me), the main character has a digital avatar embodied in his clothes that is constantly building his profile from the observation of his interactions. D-Me is essentially a very advanced personal assistant, with the autonomy of taking certain decisions in a way that resembles what Dimitrios would do in a similar situation. This way, Dimitrios is free to worry about other issues, while his digital version takes care of certain aspects for him. The third scenario details the daily life of Carmen, in a city-area AmI. Carmen can ask her AmI system to arrange for a ride for work in a shared car, asks her fridge to buy the missing ingredients for a recipe, enjoys an efficient car travelling system that avoids traffic jams and is notified when particular items she is interested in buying are on sale. Essentially, this scenario depicts the advantages of AmI in traffic management,

C. Ramos et al. (eds.), *Ambient Intelligence - Software and Applications*,
Advances in Intelligent Systems and Computing 291,
DOI: 10.1007/978-3-319-07596-9_25, © Springer International Publishing Switzerland 2014

sustainability and commerce. Finally, scenario 4 details the hypothetical use of AmI for Social Learning, with an emphasis on scheduling, people's information sharing (e.g. expertise, cv, background) and an ongoing analysis of the mental state of the participants.

Still today, all these scenarios remain in the field of science-fiction. Interestingly enough, the hardware requirements are met (e.g. we can develop a fridge that does online shopping, we can develop an app that shares our background and expertise with people around us). There are however, other questions that hold this development back. Do we really want our fridge to decide on the quality and price of the groceries we buy? Do we really want our personal information to be available to strangers? Indeed, there are many challenges that still remain and go beyond the technological ones. this paper briefly addresses these challenges and then moves on to show that, despite the challenges, the diversity and novelty of AmI applications does not decrease and, still, we expect a near future in which a more seamless integration of technology in our lives takes place.

2 What Is Holding It All Back?

It is now clear that the development of Ambient Intelligence didn't meet the initially expected pace. This can be attributed to a large (and perhaps increasing) number of challenges. One of these challenges, often disregarded by computer scientists (who form the backbone of AmI development) concerns privacy, identity and security issues. In [14] the authors make a thorough analysis of 70 AmI projects, principally in Europe, concerning these issues. They conclude that in general, current projects present a rather too sunny view of our technological future, ignoring or postponing dealing with some pressing issues. The authors also make an interesting reference to he SWAMI project (Safeguards in a World of Ambient Intelligence) which, against this trend, has constructed what they deemed "dark" scenarios [28], to show how things can go wrong in AmI and where safeguards are needed. Once again, some of these safeguards had already been put forward by [5], while others emerged more recently. As Rouvroy and Brey separately put it, the challenge here is to preserve the individual freedom to build one's own personality without excessive constrains and influences while have control over the aspects of one's identity that one projects on the world [20,6].

Marzano, on a different view, looks at the cultural implications of an unregulated or indiscriminate growth of AmI, making a parallel with the industrial revolution [1]. As later was proved to be, more was not necessarily better at the time: take for instance consequences such as the pollution. Right now, *smarter* may also not be necessarily better. Indeed, we may simply not want a smart juicer or a talking toaster. The decisions we make now will shape us as a society in the future, as the decisions in the industrial revolution resulted in today's society, for the better and the worse. We must also look at works such as [10], that point out how even the same application of AmI, in the specific case of the paper domestic assistive robots, may be perceived differently, and must thus be developed differently, in different cultures, even when these different cultures are within the European context.

Then, and moving to what is most likely the field of the reader, one must also consider the immense technological challenges that are still ahead. One the one hand, we have the challenges that are related to the physical constraints and nature of the necessary hardware. In [12] the authors examine the intricate relationship between the growing need for more computational and communicational power to support increasingly complex services and, at the same time, the need for smaller, more lightweight and efficient devices. It is easy to understand how the objectives of these two fields conflict. Many other technological challenges exist. [8] provide a fairly detailed analysis of these challenges, touching issues such as: (1) the need for increase sensibility, closely related to the assessment of needs and preferences (e.g. a system should be able to know when a new event is important enough to interrupt a user at a given time); (2) the limits of battery life; (3) the simultaneous modelling of several users in the same environment; (4) methods to increase human socialization and interaction in order to avoid the risk of encouraging people to stay home enjoying their AmI systems; (5) the need for increased autonomy in middleware layers; among many others.

Another issue holding back a faster development of AmI that is not so frequently mentioned is the scatter of research efforts. Indeed there are currently many different institutions doing research on very similar topics. When these institutions want to conciliate efforts they may find it difficult to do so since they use different technologies, standards or approaches. Facilitating this integration and interoperability could result in a coming closer of different teams, whom could join efforts and more efficiently work together for the same goal.

3 Ambient Intelligence: Diversity and Innovation

Despite the (non-exhaustive) list of challenges presented in the previous section, which seems to only get bigger as research projects approach implementation phases, new fields of application, with extensive lists of advantages, continue to emerge. These prospective advantages are what keeps pushing research forward, despite the numerous challenges. This section presents several fields in which innovative applications of AmI are being developed.

Context-Acquisition

The gathering, storing, management and provision of contextual information about the user for supporting decision-making are some of the key operations in most AmI systems. Indeed, and depending on the domain and objectives of the application, aspects such as the geographical location, the time of the day, the level of noise or luminosity, the task the user is currently enrolled in or the persons the user is with may be of preponderant importance for the correct provision of services. Particularly, a growing interest exists in the scientific community towards the development of approaches that can acquire and mange context information autonomously and with minimum explicit interaction with the user.

Examples of such approaches can be found in [18], in which the conflict handling style of users of an Online Dispute Resolution tool is automatically classified, or in [25], in which several examples of context-awareness in mobile tourism guides are presented. More general-purpose approaches can also be found, such as CoWSAMI, a middleware infrastructure that enables context awareness in open ambient intelligence environments, consisting of mobile users and context sources that become dynamically available as the users move from one location to another [3].

Education

The technological evolution led to significant changes in the Teacher-Student relationship, which does not take part necessarily face-to-face any more and may rely on communication tools that are always poor replacements for our rich communication processes. Interestingly enough, AmI seems to be one possible way of dealing with some of the unaddressed challenges that result from this rapid (and sometimes without control) technological change. One of these challenges is the lack of contextual information of current online communication tools, which are often regarded as cold and impersonal. Teachers and students, that until recently communicated face-to-face, now do it over forums and text-based chats, which makes communication far poorer. In [19], the authors depict a non-invasive way of building contextual information about users of e-Learning tools, that can compile important information for improving the decision-making processes of teachers, namely concerning aspects such as the level of stress or fatigue of students. To overcome similar issues, [17] developed virtual tutoring anthropomorphic characters that provide guidance and motivation to the users and give the feeling that users are being observed, followed and their actions understood. To better accomplish this, the characters are enriched with contextual behavioural models. On another example [2] detail an augmented desk, to be used in the classroom, and improve the learning experience of the student. In a more specific

Robotics

Robots were always part of the Artificial Intelligence imaginary. Traditionally, each person, in a near future, would presumably have a robotic companion, probably in a humanoid form, taking the role of a multi-faceted personal assistant that would cater to all our needs. Research and technological evolution seem to be pointing to other direction. Indeed, the most likely scenario is the one in which an ecology of agents, generally our traditional appliances and hardware, have communication capabilities and coordinate in order to service us, through some kind of enabling middleware [13]. Robots, literally, can also be conceived that behave similarly, sharing our environment and moving naturally through it while providing their services [23]. A mixed scenario is also envisioned by researchers in which traditional appliances and robots, in the true sense of the word, coordinate their efforts in order to accomplish their goals and, in doing so, implement AmI [22].

Health

The Healthcare sector is currently one of the technologically most advanced ones. It thus comes as no surprise that many different projects can be found in this field. Two main trends can be identified: the provision of healthcare services at home and the support of healthcare in the hospital. In both scenarios, the main aim is to decrease healthcare costs while making it more accessible, personalized and available [21]. In the home scenario, most of the projects target the support of the elderly, the so-called assisted living, to provide them with more independence and quality of life [7]. In the healthcare institution, one can find projects to deal with very specific types of scenarios (e.g. Alzheimer patients [11]) as well as more generic ones such as the use of discrete hidden Markov models for classifying the activities of healthcare practitioners [24]. Concerning the very particular case of mental disabilities, [26] presents a tool for the assessment of cognitive skills of mentally disabled working people, which can be quite important for improving disabled people's performance and self-esteem in their workplaces.

Transportation

Given that we spend a significant part of our lives travelling back and forth, transportation is another field in which the use of Ambient Intelligence could translate into multiple advantages, as pointed out by [8]: (1) train stations, buses, and cars can be equipped with technology that can provide fundamental knowledge about how the system is performing; (2) identification of potential improvements by using the system more effectively, that can result in the improvement of the experience of people; (3) GPS-based spatial location of public transportation vehicles and services; (4) vehicle identification and image processing to make transport more fluent and hence more efficient and safe, or to detect situations of interest in busy conditions [27], among many others. The coming together of these and other technological evolutions could result in significant improvements in currently challenged transportation networks.

Emergency Services

Emergency services and their organization constitute a complex and critical field in which one can also find applications for Ambient Intelligence. Here, the main objectives are the provision of mission-critical knowledge in real-time and the coordination of the different actors, as addressed by the AMIRA project [4]. [16] further discuss the possibilities of AmI in emergency services, namely their potential role in the rapid organization of virtual response teams as well as the support for the communication and coordination of the members of such teams, even in the cases in which they span different countries.

Culture

Ambient Intelligence is also being researched as a way of implementing better access to sources of culture. An example of a project in a domestic environment is GENIO, that besides other services provides the user with easy and intuitive access to music, videos and pictures, accessible through voice commands "heard" by the system from a microphone worn by the user on the chest [15]. In an outdoors example one can find DALICA: a project that provided access to information about geographically spread monuments in a geographical area in Villa Adriana, Italy [9]. The users can visit the different stone monuments and automatically receive information on the smartphone as they approach them. In the overall, AmI can make access to culture more intuitive, personalized and automatized, making the whole experience more pleasant.

4 Conclusions

Ambient Intelligence has not grown as fast as initially expected. The natural interfaces, always present and proactive assistants or the seamless experience still seem far from reality. Several challenges have been pointed out that can explain, at least in part, this delay. These include technological (mostly at the level of integration), cultural, and safety and privacy-related challenges.

Interestingly enough, this does not prevent new ideas and new fields of application from emerging. Indeed, we continue to see an increasing diversity of the areas in which AmI would result in amazing advantages. These areas include, as depicted in this paper, sensitive ones such as healthcare, transportation or emergency services as well as more "traditional" ones such as education or home assistants.

It seems rather contradictory that a field with so many challenges still to be addressed continues to expand, to grow in different directions, most likely finding additional new challenges. The truth is that this expansion cannot stop. The real possibilities of AmI and the advantages of its application in so many fields are what continues to push research forward. There is the need to show what can actually be done once the challenges are overcome. Once these challenges are overcome and the technology finally unravels as expected, what was once science fiction will become reality, giving the feeling of a new technological evolution.

Acknowledgements. This work is part-funded by ERDF - European Regional Development Fund through the COMPETE Programme (operational programme for competitiveness) and by National Funds through the FCT - Fundação para a Ciência e a Tecnologia (Portuguese Foundation for Science and Technology) within project FCOMP-01-0124-FEDER-028980 (PTDC/EEI-SII/1386/2012). This work is part-funded by National Funds through the FCT - Fundação para a Ciência e a Tecnologia (Portuguese Foundation for Science and Technology) within projects PEst-OE/EEI/UI0752/2011.

References

1. Aarts, E.H., Marzano, S.: The new everyday: Views on ambient intelligence (2003)
2. Antona, M., Margetis, G., Ntoa, S., Leonidis, A., Korozi, M., Paparoulis, G., Stephanidis, C.: Ambient Intelligence in the classroom: an augmented school desk. In: Proceedings of the 2010 AHFE International Conference, pp. 17–20 (2010)
3. Athanasopoulos, D., Zarras, A.V., Issarny, V., Pitoura, E., Vassiliadis, P.: CoWSAMI: Interface-aware context gathering in ambient intelligence environments. Pervasive and Mobile Computing 4(3), 360–389 (2008)
4. Bergmann, R.: Ambient intelligence for decision making in fire service organizations. In: Schiele, B., Dey, A.K., Gellersen, H., de Ruyter, B., Tscheligi, M., Wichert, R., Aarts, E., Buchmann, A.P. (eds.) AmI 2007. LNCS, vol. 4794, pp. 73–90. Springer, Heidelberg (2007)
5. Bogdanowicz, M., Scapolo, F., Leijten, J., Burgelman, J.C.: Scenarios for ambient intelligence in 2010, pp. 3–8. Office for official publications of the European Communities (2001)
6. Brey, P.: Freedom and privacy in ambient intelligence. Ethics and Information Technology 7(3), 157–166 (2005)
7. Cesta, A., Cortellessa, G., Pecora, F., Rasconi, R.: Exploiting scheduling techniques to monitor the execution of domestic activities. Intelligenza Artificiale 2, 4 (2005)
8. Cook, D.J., Augusto, J.C., Jakkula, V.R.: Ambient intelligence: Technologies, applications, and opportunities. Pervasive and Mobile Computing 5(4), 277–298 (2009)
9. Costantini, S., Mostarda, L., Tocchio, A., Tsintza, P.: DALICA: Agent-based ambient intelligence for cultural-heritage scenarios. Intelligent Systems, IEEE 23(2), 34–41 (2008)
10. Cortellessa, G., Scopelliti, M., Tiberio, L., Svedberg, G.K., Loutfi, A., Pecora, F.: A cross-cultural evaluation of domestic assistive robots. In: Proceedings of the AAAI Fall Symposium on AI and Eldercare (2008)
11. Corchado, J.M., Bajo, J., Abraham, A.: GerAmi: Improving healthcare delivery in geriatric residences. Intelligent Systems, IEEE 23(2), 19–25 (2008)
12. De Man, H.: Ambient intelligence: gigascale dreams and nanoscale realities. In: IEEE International Solid-State Circuits Conference. Digest of Technical Papers. ISSCC, pp. 29–35. IEEE (2005)
13. Encarnação, J.L., Kirste, T.: Ambient intelligence: Towards smart appliance ensembles. In: Hemmje, M., Niederée, C., Risse, T. (eds.) From Integrated Publication and Information Systems to Information and Knowledge Environments. LNCS, vol. 3379, pp. 261–270. Springer, Heidelberg (2005)
14. Friedewald, M., Vildjiounaite, E., Punie, Y., Wright, D.: The Brave New World of Ambient Intelligence: An Analysis of Scenarios Regarding Privacy, Identity and Security Issues. In: Clark, J.A., Paige, R.F., Polack, F.A.C., Brooke, P.J., et al. (eds.) SPC 2006. LNCS, vol. 3934, pp. 119–133. Springer, Heidelberg (2006)
15. Gárate, A., Herrasti, N., López, A.: GENIO: an ambient intelligence application in home automation and entertainment environment. In: Proceedings of the 2005 Joint Conference on Smart Objects and Ambient Intelligence: Innovative Context-Aware Services: Usages and Technologies, pp. 241–245. ACM (2005)
16. Jones, V., Karagiannis, G., Heemstra de Groot, S.: Ad hoc networking and ambient intelligence to support future disaster response (2005)

17. Ndiaye, A., Gebhard, P., Kipp, M., Klesen, M., Schneider, M., Wahlster, W.: Ambient intelligence in edutainment: Tangible interaction with life-like exhibit guides. In: Maybury, M., Stock, O., Wahlster, W. (eds.) INTETAIN 2005. LNCS (LNAI), vol. 3814, pp. 104–113. Springer, Heidelberg (2005)
18. Novais, P., Carneiro, D., Gomes, M., Neves, J.: Behavioural and Context Analysis in an Online Dispute Resolution Environment. In: Proceedings of the Sixth International Workshop on Juris-Informatics (JURISIN 2012), hosted by the Fourth JSAI International Symposia on AI (2012)
19. Rodrigues, M., Gonçalves, S., Carneiro, D., Novais, P., Fdez-Riverola, F.: Keystrokes and clicks: Measuring stress on E-learning students. In: Casillas, J., Martínez-López, F.J., Vicari, R., De la Prieta, F. (eds.) Management Intelligent Systems. Advances in Intelligent Systems and Computing, vol. 220, pp. 119–126. Springer, Heidelberg (2013)
20. Rouvroy, A.: Privacy, data protection, and the unprecedented challenges of ambient intelligence. Studies in Ethics, Law, and Technology 2(1) (2008)
21. Sadri, F.: Ambient intelligence: A survey. ACM Computing Surveys (CSUR) 43(4), 36 (2011)
22. Saffiotti, A., Broxvall, M.: PEIS ecologies: Ambient intelligence meets autonomous robotics. In: Proceedings of the 2005 Joint Conference on Smart Objects and Ambient Intelligence: Innovative Context-Aware Services: Usages and Technologies, pp. 277–281. ACM (2005)
23. Sanfeliu, A., Hagita, N., Saffiotti, A.: Network robot systems. Robotics and Autonomous Systems 56(10), 793–797 (2008)
24. Sánchez, D., Tentori, M., Favela, J.: Activity recognition for the smart hospital. IEEE Intelligent Systems 23(2), 50–57 (2008)
25. Schwinger, W., Grün, C., Pröll, B., Retschitzegger, W., Schauerhuber, A.: Context-awareness in mobile tourism guides A comprehensive survey. Rapport Technique. Johannes Kepler University Linz (2005)
26. Vilaro, A., Orero, P.: User-centric cognitive assessment. Evaluation of attention in special working centres: from paper to Kinect. Advances in Distributed Computing and Artificial Intelligence Journal 1(7), Special Issue 7 (2013)
27. Velastin, S.A., Boghossian, B.A., Lo, B.P., Sun, J., Vicencio-Silva, M.: PRISMATICA: toward ambient intelligence in public transport environments. IEEE Transactions on Systems, Man and Cybernetics, Part A: Systems and Humans 35(1), 164–182 (2005)
28. Wright, D.: The dark side of ambient intelligence. Info 7(6), 33–51 (2005)

AmI: Monitoring Physical Activity

Ricardo Costa[1], Luís Calçada[1], Diva Jesus[2], Luís Lima[3], and Luís C. Lima[1]

[1] CIICESI, College of Technology and Management of Polytechnic of Porto, Felgueiras, Portugal
{rcosta,8070032,llima}@estgf.ipp.pt
[2] PM&R Medical Resident, Hospital Central do Funchal, Funchal, Portugal
[3] PM&R Medical Resident, Centro Hospitalar Tondela Viseu, Viseu, Portugal

Abstract. With the increase of the Information and Communication Technologies in all aspects and sectors of our life, a world of unprecedented scenarios is arising. Some of these scenarios are the explosions of new and different Healthcare projects that intend to decrease the economical and social costs of the real ageing population phenomenon, through the delocalization of healthcare services delivery and management to the home. These Ambient Assisted Living environments, which we have taken a step forward with the introduction of proactive techniques for better adapting to its users, usually elderly or chronic patients, are now entering a new phase: users with no special help needs also need to have a healthier Active Ageing. Since the direct correlation between physical activity and health is well established, we present here an idea to better achieve this goal.

Keywords: Ambient Assisted Living, e-Health, Physical Monitor, Active Ageing.

1 Introduction

The direct correlation between physical activity and health is well established by numerous publications [1]. Also, these reports called attention to the health-related benefits of regular physical activity that did not meet traditional criteria for improving fitness levels - for example, short sessions (less than 20 minutes) and less than 50% of aerobic capacity. An important goal was to clarify for the public and health professionals the amounts and intensities of physical activity needed for improved health, lowered susceptibility to disease, and decreased mortality. Moreover, these publications documented the dose-response relationship between physical activity and health: some activity is better than none, and more activity, up to a point, is better than less. It is clear that additional amounts of physical activity provide additional health benefits. There is also evidence for an inverse dose-response relationship between physical activity and all-cause mortality, overweight, obesity and fat distribution, type 2 diabetes, colon cancer, quality of life, and independent living in older adults.

Two conclusions from the 1996 U. S. Surgeon General's Report, presented in [1], are worth to note: i) important health benefits can be obtained by including a

C. Ramos et al. (eds.), *Ambient Intelligence - Software and Applications*,
Advances in Intelligent Systems and Computing 291,
DOI: 10.1007/978-3-319-07596-9_26,

moderate amount of physical activity on most, if not all, days of the week; and, ii) additional health benefits result from greater amounts of physical activity.

A diversity of exercises to improve the components of physical fitness is recommended for all adults. The health-related components of physical fitness include cardiovascular (aerobic) fitness, muscular strength and endurance, flexibility, and body composition. Other exercises improving neuromuscular fitness, for example balance and agility, are also recommended, principally for older adults.

Various methods are used to guide exercise prescription, including Heart Rate, Oxygen Uptake, Ratings of Perceived Exertion (RPE), OMNI, talk test, affective valence, absolute energy expenditure per minute (kcal·min-1), percentage-predicted maximum HR (HRmax), percent oxygen update, and METs (metabolic equivalents) [1]. Each of these methods for monitoring prescribed exercise may result in health/fitness improvements when properly applied [1].

In the context of VirtualECare, a high level of interaction and user feedback response to the prescribed exercise can be achieved using RPE method. The 6-20 Borg Rating of Perceived Exertion [2] is a way of measuring physical activity intensity level. Perceived exertion is how hard you feel like your body is working. It is based on the physical sensations a person experiences during physical activity, including increased heart rate, increased respiration or breathing rate, increased sweating, and muscle fatigue. As an alternative to using HR alone to clinically determine intensity of exercise, the Borg rating is useful [3]. Monitoring of physical activity is also important in so many contexts [4, 5]. For example, the inability to perform exercise without discomfort may be one of the first symptoms experienced by patients with heart failure and is often the principal reason for seeking medical care [6].

Recording the Borg rating is easily implemented with VirtualECare and may serve several objectives:

- As a way to prescribe physical exercise;
- To monitor the physical activity, compare it with prescribed exercises in order to adjust their intensity;
- To monitor a health status that may require medical attention.

1.1 Related Work

Although the use of technology for monitoring health at home is now widespread, some challenges are yet unexplored, like the integration of common home appliances (TV) and mobile platforms (smartphones, tablets) with information technologies.

The availability of smartphones equipped with a rich set of sensors has enabled ubiquitous human activity recognition on mobile platforms. Monitoring daily physical activities and their levels of intensity helps in recognizing the health and wellness of the users as a real-world application. Mobile phones are ubiquitous, unobtrusive, and easy to use, making them a suitable platform for inducing behaviour change for a healthier and more active lifestyle.

Several health related strategies and types of interventions have been implemented with mobile phones [7]: 1) tracking health information; 2) involving the healthcare team; 3) leveraging social influence; 4) increasing the accessibility of health information; 5) utilizing entertainment. The first two are of particular interest for VirtualECare. Tracking health information and involving the healthcare team has been achieved using a variety of intervention types, such as, but not limited to [7]:

- Native applications designed to support recording one or more health-related behaviors (e.g., physical activity or food intake), and relevant measures (e.g., blood glucose levels, or blood pressure), the most interesting ones implementing automated sensing for tracking;
- Remote coaching interventions - the tracking data collected on mobile phones is uploaded to a website where it is reviewed by a member of the healthcare team, who works with patients to help them learn to manage their conditions more effectively;
- In addition to remote coaching, mobile phones are frequently used to monitor patient's health and to alert the healthcare team if risky symptoms develop

Modern smartphones already offer media-rich and context-aware features that are useful for e-health applications. A 2011 review for the iOS platform alone [8] identified more than 200 health-related applications. Although the biggest group was about medical information reference, the mobile users largely favoured tracking tools.

Specifically for the elderly, mobile phones are promising functionalities and applications that can satisfy the requirements and needs of older people, in order to improve their quality of life. The needs and expectations of the older persons adopting mobile phone include [9]: feeling safe and secure; memory and daily life activity aids; traditional use as communication device enabling contacts with friends and family; freedom of movement (involves both self-determination and empowerment); and services that promote their physical and mental well-being for a healthier independent life.

Many publications show that mobile phones are fit for activity recognition for everyday life and physical activity monitoring using built-in sensors [10-15]. In the same line of thought, more elaborated uses challenge our attention, such as gesture recognition [16], prescribing and monitoring rehabilitation exercises [17] or the use of serious games to promote health [18, 19].

1.2 VirtualECare

The VirtualECare project (Figure 1) envisions a new and effective way of providing healthcare, where the treatment will no more be institution centred but, instead, will shift to be user centred, resulting in a better and cheaper service for both the user and the provider, through remote monitoring and assessment of user vital data and real time location. This same data, after adequate processing, can conduct to alarms, recommendations or even actions to assist the needed patient while he maintains is

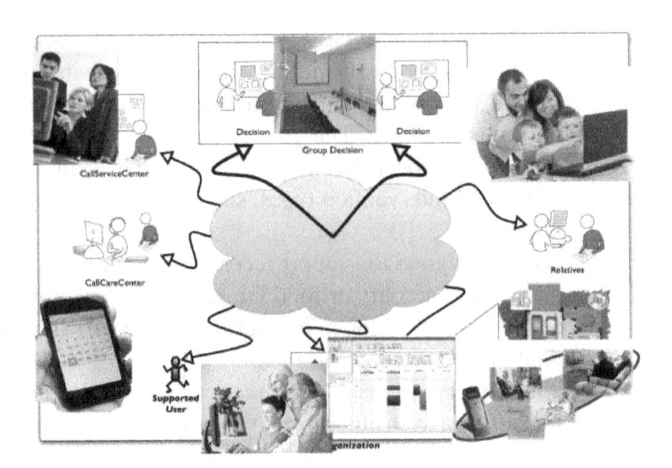

Fig. 1. The Original VirtualECare Project

normal, day-to-day, life. These new healthcare provision approximations bring great advantages that may also be extended to relatives and friends of the patient, since they can be informed, in real time, of his actual condition allowing them to also intervene in case of necessity [20, 21].

As a consequence of this new way of healthcare provision we had the need to enrich the patient home with Ambient Intelligence (AmI) and, more concretely, Ambient Assisted Living (AAL) technologies. This approximation allowed us to better respond to the patient specific needs, modulating his home environment as needed in order to better respond to his expectations [22].

2 VirtualECare2

This project aims to develop a new module to the VirtualECare platform [23], that is already able to remotely monitor the health and wellness of its users while comfortably in their usual habitat (at home), allowing the formal caregivers and relatives to closely follow their condition and evolution. Additionally, this also represents the first step in the evolution to VirtualECare2 with new user, non-obstructive, interfaces, inclusive in the used hardware, based on TV and mobile platforms (e.g. tablets and smartphones). The objective is to use hardware that is already, and each day more traditionally, inside, and even considered part, of the user traditional habitat (e.g. home). We believe that through the use of this type of hardware the usual user initial renitence will be easily overtaken.

2.1 Physical Activity Monitor

Currently we are planning to use the Physical Activity Monitor module not only to monitor the user day-to-day activity, but also to suggest essential wellbeing exercises in order to help maintaining some basic activity. Additionally, and in case of users with prescribed exercises, the Physical Activity Monitor may also be used to doctor interaction, supervision and progress report.

All the user interaction with the Physical Activity Monitor will be based on VirtualECare2 new user-friendly interface, based on TV with Android operating system, including Google TV/Android TV. This kind of interaction, using traditional hardware, is known for best and fastest adoption by users. The physical movements monitoring will be made using Android based sensors smartphones, simulating the more traditional Kinetic approach, but without the need for the Kinetic hardware.

Since VirtualECare system architecture (Figure 3) was idealized to be a modular one, in order to allow the addition of new modules in the future, the addition of this module is easily done using the already existing underlying OSGi framework, inside the user home, since the Physical Activity Monitor should not stand in the "cloud", but in the user traditional environment, in some way similar to the Recognition Module [24], interconnected with the existing local technologies. This decision will also allow user monitored Physical Activity even if, for some reason, the connection to the "cloud" is lost, being the events afterwards synchronized recurring to R-OSGi or the more traditional web services approach.

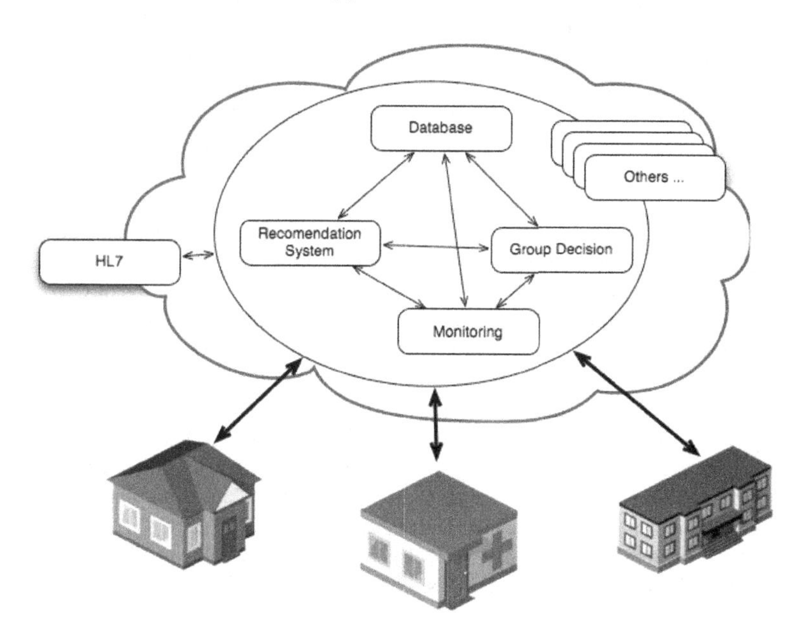

Fig. 2. VirtualECare Modular Architecture

3 Conclusions and Future Work

In this paper we present the initial evolution for the original VirtualECare platform to VirtualECare2 platform. Although maintaining the user-centric initial characteristics, in this version 2 approach we pretend to take them to a all new level of human computer interaction. For the first version 2 launch we pretend to add one additional module, the Physical Activity Module, build with these new interaction and them, progressively, migrate all the other modules. With the Physical Activity Monitor we are bringing the use o TV and smartphone hardware to the user environments, or better, we are giving the already available TV and smartphone some additional usage. We believe users will better accept the introduction of such additional usage since no new, and somehow "strange" hardware, is added to the environment.

We are now working on the development of this new-presented module and verifying how it can be seamless integrated in our AAL system. We are also testing some low cost TV devices (e.g. Google TV and Android TV) and smartphone interaction, which will allow us to achieve the pretended results.

References

1. Ehrman, J.K.: ACSM's Resource Manual for Guidelines for Exercise Testing and Prescription. Wolters Kluwer Health/Lippincott Williams & Wilkins, Philadelphia (2010)
2. Borg, G.A.: Psychophysical bases of perceived exertion. Medicine and Science in Sports and Exercise 14, 377–381 (1982)
3. Fletcher, G.F., Balady, G.J., Amsterdam, E.A., Chaitman, B., Eckel, R., Fleg, J., Froelicher, V.F., Leon, A.S., Piña, I.L., Rodney, R., Simons-Morton, D.A., Williams, M.A., Bazzarre, T.: Exercise Standards for Testing and Training: A Statement for Healthcare Professionals From the American Heart Association. Circulation 104, 1694–1740 (2001)
4. Mezzani, A., Hamm, L.F., Jones, A.M., McBride, P.E., Moholdt, T., Stone, J.A., Urhausen, A., Williams, M.A.: Aerobic exercise intensity assessment and prescription in cardiac rehabilitation: a joint position statement of the European Association for Cardiovascular Prevention and Rehabilitation, the American Association of Cardiovascular and Pulmonary Rehabilitation and the Canadian Association of Cardiac Rehabilitation. European Journal of Preventive Cardiology 20,442 467, 442–467 (2013)
5. Piepoli, M.F., Conraads, V., Corrà, U., Dickstein, K., Francis, D.P., Jaarsma, T., McMurray, J., Pieske, B., Piotrowicz, E., Schmid, J.-P., Anker, S.D., Solal, A.C., Filippatos, G.S., Hoes, A.W., Gielen, S., Giannuzzi, P., Ponikowski, P.P.: Exercise training in heart failure: from theory to practice. A consensus document of the Heart Failure Association and the European Association for Cardiovascular Prevention and Rehabilitation. European Journal of Heart Failure 13, 347–357 (2011)
6. Pina, I.L., Apstein, C.S., Balady, G.J., Belardinelli, R., Chaitman, B.R., Duscha, B.D., Fletcher, B.J., Fleg, J.L., Myers, J.N., Sullivan, M.J.: Exercise and heart failure: A statement from the American Heart Association Committee on exercise, rehabilitation, and prevention. Circulation 107, 1210–1225 (2003)
7. Klasnja, P., Pratt, W.: Healthcare in the pocket: mapping the space of mobile-phone health interventions. Journal of Biomedical Informatics 45, 184–198 (2012)
8. Liu, C., Zhu, Q., Holroyd, K.A., Seng, E.K.: Status and trends of mobile-health applications for iOS devices: A developer's perspective. Journal of Systems and Software 84, 2022–2033 (2011)
9. Plaza, I., Martín, L., Martin, S., Medrano, C.: Mobile applications in an aging society: Status and trends. Journal of Systems and Software 84, 1977–1988 (2011)

10. Bieber, G., Voskamp, J., Urban, B.: Activity Recognition for Everyday Life on Mobile Phones. In: Stephanidis, C. (ed.) UAHCI 2009, Part II. LNCS, vol. 5615, pp. 289–296. Springer, Heidelberg (2009)
11. Hynes, M., Wang, H., McCarrick, E., Kilmartin, L.: Accurate monitoring of human physical activity levels for medical diagnosis and monitoring using off-the-shelf cellular handsets. Pers Ubiquit Comput 15, 667–678 (2011)
12. Incel, O., Kose, M., Ersoy, C.: A Review and Taxonomy of Activity Recognition on Mobile Phones. BioNanoSci 3, 145–171 (2013)
13. Ouchi, K., Doi, M.: A Real-Time Living Activity Recognition System Using Off-the-Shelf Sensors on a Mobile Phone. In: Beigl, M., Christiansen, H., Roth-Berghofer, T.R., Kofod-Petersen, A., Coventry, K.R., Schmidtke, H.R. (eds.) CONTEXT 2011. LNCS, vol. 6967, pp. 226–232. Springer, Heidelberg (2011)
14. Ouchi, K., Doi, M.: Living activity recognition using off-the-shelf sensors on mobile phones. Ann. Telecommun. 67, 387–395 (2012)
15. Sun, L., Zhang, D., Li, N.: Physical Activity Monitoring with Mobile Phones. In: Abdulrazak, B., Giroux, S., Bouchard, B., Pigot, H., Mokhtari, M. (eds.) ICOST 2011. LNCS, vol. 6719, pp. 104–111. Springer, Heidelberg (2011)
16. Wang, X., Tarrío, P., Metola, E., Bernardos, A.M., Casar, J.R.: Gesture Recognition Using Mobile Phone's Inertial Sensors. In: Omatu, S., Paz Santana, J.F., González, S.R., Molina, J.M., Bernardos, A.M., Rodríguez, J.M.C. (eds.) Distributed Computing and Artificial Intelligence. AISC, vol. 151, pp. 173–184. Springer, Heidelberg (2012)
17. Goodney, A., Jung, J., Needham, S., Poduri, S.: Dr. Droid: Assisting Stroke Rehabilitation Using Mobile Phones. In: Gris, M., Yang, G. (eds.) MobiCASE 2010. Lecture Notes of the Institute for Computer Sciences, Social Informatics and Telecommunications Engineering, vol. 76, pp. 231–242. Springer, Heidelberg (2012)
18. Smith, S.T., Talaei-Khoei, A., Ray, M., Ray, P.: Agent-Based Monitoring of Functional Rehabilitation Using Video Games. In: Brahnam, S., Jain, L.C. (eds.) Advanced Computational Intelligence Paradigms in Healthcare 5. SCI, vol. 326, pp. 113–141. Springer, Heidelberg (2010)
19. Wattanasoontorn, V., Hernández, R., Sbert, M.: Serious Games for e-Health Care. In: Cai, Y., Goei, S.L. (eds.) Simulations, Serious Games and Their Applications, pp. 127–146. Springer, Heidelberg (2014)
20. Costa, R., Novais, P., Lima, L., Carneiro, D., Samico, D., Oliveira, J., Machado, J., Neves, J.: VirtualECare: Intelligent Assisted Living. Electronic Healthcare, vol. 1, pp. 138–144. Springer, Heidelberg (2009)
21. Costa, R., Novais, P., Lima, L., Cruz, J.B., Neves, J.: VirtualECare: Group Support in Collaborative Networks Organizations for Digital Homecare, pp. 151–178 (2009)
22. Carneiro, D., Costa, R., Novais, P., Neves, J., Machado, J., Neves, J.: Simulating and Monitoring Ambient Assisted Living. In: ESM 2008, pp. 175–182 (Year)
23. Costa, R., Novais, P., Lima, L., Carneiro, D., Samico, D., Oliveira, J., Machado, J., Neves, J.: VirtualECare: Intelligent Assisted Living. In: Weerasinghe, D. (ed.) eHealth 2008. LNICST 1, pp. 138–144. Springer, Heidelberg (2009)
24. Almeida, A., Costa, R., Lima, L., Novais, P.: Non-obstructive Authentication in AAL Environments. In: 7th International Conference on Intelligent Environments (IE 2011), pp. 63–73. IOS Press (Year)

Multi-agent Technology to Perform Odor Classification

Sigeru Omatu[1], Tatsuyuki Wada[1], Sara Rodríguez[2],
Pablo Chamoso[2], and Juan M. Corchado[2]

[1] Osaka Institute of Technology, Osaka, Japan
[2] Computer and Automation Department, University of Salamanca, Spain
omatu@rsh.oit.ac.jp, coco.tk.family@gmail.com,
{srg,chamoso,corchado}@usal.es

Abstract. Quartz crystal microbalance (QCM) sensors are used to measure and classify odors. In this paper, we use seven QCM sensors and three kinds of odors. The system has been developed as a virtual organization of agents using the agent platform called PANGEA (Platform for Automatic coNstruction of orGanizations of intElligents Agents), which is a platform to develop open multi-agent systems, specifically those including organizational aspects. The main reason that justifies the use of the agents is the scalability of the platform; that is, the way in which it models the services. The functionalities of the system are modeled as services inside the agents, or as SOA (Service Oriented Approach) architecture compliant services using Web Services. In this way, it is possible to improve odor classification systems with new algorithms, tools and classification techniques.

Keywords: Odor sensing, odor classification, multi-agent systems, virtual organizations, QCM sensors.

1 Introduction

During the last years, major advances have been made in the field of Ambient Intelligence [1], [2], which has come to acquire significant relevance in the daily lives of people [5], [6], [7]. Ambient Intelligence adapts technology to people's needs by proposing 3 concepts: ubiquitous computing, ubiquitous communication and intelligent user interfaces. The development of new frameworks and models to allow information access, independently of the location, is needed in order to achieve these targets. Wireless sensor networks [3], [4], [22], provide an infrastructure, which is able to distribute communications in dynamic environments by incrementing mobility and efficiency independently of the location. Sensor networks interconnect a large amount of sensors and manage information in the intelligent environment. Many times information management is done in a distributed way. However, it is necessary to have distributed systems with enough capabilities to manage sensor networks in an efficient way and to include elements with some degree of intelligence that can be embedded in the devices and act both autonomously and in coordination with the distributed system. Multi-agent systems are a suitable alternative to perform this type of systems.

C. Ramos et al. (eds.), *Ambient Intelligence - Software and Applications*,
Advances in Intelligent Systems and Computing 291,
DOI: 10.1007/978-3-319-07596-9_27, © Springer International Publishing Switzerland 2014

There are several proposals to build smart environments that combine multi-agent systems and sensor networks [8], [9], [10], [11], [12], [13] , [14], [15], [16], [17], [18], [19], [20], [21]. New approaches are needed to support evolutional systems and to facilitate their growth and runtime updates. The dynamics of open environments have promoted the use of Virtual Organizations of Agents (VOs). A VO [25], [26], [27], [28], [29] is an open system designed for grouping; it allows for the collaboration of heterogeneous entities and provides a separation between the form and function that define their behavior. However, it is not possible to find an existing multi-agent architecture to work on the concept of virtual organizations and to provide agents capable of working with any type of sensor or device. This article considers different types of odor sensors and aims to classify odors according to sensing data by using quartz crystal microbalance (QCM) sensors. QCM sensors are sensitive to odors and allow the precise measurement of odor data. Using many QCM sensors, we will attempt to classify various kinds of odors based on neural networks. To model the system, virtual organizations of agents, which are capable of bringing a greater number of possibilities, are presented. These agents are connected with PANGEA [23], a multi-agent platform designed on the basis of virtual organizations, aimed at the creation of intelligent environments.

Over the last decade, odor-sensing systems (called electronic nose (EN) systems) have undergone important developments from a technical and commercial point of view. EN refers to the ability to reproduce the human sense of smell by using sensor arrays and pattern recognition systems [30].

The authors in [31] present a type of an EN system to classify various odors under the various densities of odors based on a competitive neural network by using learning vector quantization (LVQ). The odor data were measured by an odor sensor array made of MOGSs. We used fourteen MOGSs of FIGARO Technology Ltd in Japan. We considered two types of data for classification in the experiment. The first type included four types of teas, while the second included five types of coffees with similar properties. The classification results of teas and coffees were approximately 96% and 89% respectively, which was much better than the results in [32], [24].

The article is structured as follows. First, the PANGEA platform is described in section 2, detailing the structure of the virtual organizations used in the odor classification case study. Both the platform and virtual organizations are evaluated in a case study consisting of an intelligent environment for odor recognition. Finally the results of the case study and the conclusions reached from this research are presented.

2 Case Study: Development of a VO for Odor Classification

This central section of the article presents the integration of the system and the sensors used in the multi-agent architecture, and explains the main concepts of QCM sensors. In addition, an overview of the odor sensing system and the measures of odor data used are described.

2.1 Integration in a Multi-agent Platform (PANGEA)

With the development of ubiquitous and distributed systems, it is interesting to have new agent platforms that facilitate the development of open agent-architectures that can be deployed on any device. PANGEA [23] is an agent platform based on organizational concepts. It can model and implement all kinds of open systems, encouraging the sharing of resources and facilitating control of all nodes where the different agents are deployed.

It is essential to have control mechanisms that enable new devices to be included in a single platform where they can be easily integrated, managed and monitored. In this case PANGEA, with its model of agents and organizations, provides the necessary features to function as the base platform when developing a comprehensive system.

In order to facilitate control of the organization, PANGEA has several agents that are automatically deployed when starting the platform operation: OrganizationManager and OrganizationAgent are in charge of the management of the organizations and suborganizations; InformationAgent is in charge of accessing the database containing all pertinent system information; ServiceAgent is in charge of recording and controlling the operation of services offered by the agents; NormAgent is in charge of the norms in the organization; and CommunicationAgent is in charge of controlling communication among agents, and recording the interaction between agents and organizations.

In addition to the intrinsic PANGEA agents, the organizations developed in the present system are the following:

- *Odor-recognition sensors organization.* In this organization all agents belonging to an individual odor recognition system is deployed. Such agents may also be of different types (sensor agents, interface agent and identifier agent).
- *Sensor control central organization.* In this organization the agent interface type is included, representing each of the odor-recognition sensors organizations together with an adapter agent.

Communication in this case is restricted only to the existing agents in the same organization, in addition to the control agents that the PANGEA platform offers (as is the case of the Information Agent, which accesses the database).

Each type of agent is engaged in a well-defined task, as explained below:

- *Sensor agent.* It is exclusively dedicated to performing sensor readings and providing the latest value when an authorized agent requires such data.
- *Identifier agent.* Its function is to perform the necessary calculations for the identification of odors. It makes use of the ability to communicate with the sensor agents, which require the data needed to perform these calculations.
- *Interface agent.* This kind of agent is present in the two types of virtual organizations cited. It is responsible for providing a communication link with the agents outside their own organization of odor-recognition sensors that are authorized to establish two-way communications using the appropriate communication format.

- *Classifier agent.* This agent performs classification services that implement the algorithms described in the following section. To perform the classification, a Classification Method of Odor Data (e-BPNN) is used and implemented on the platform. We can say that the classifier agent could use new methods for the classification of odors by making the system scalable in terms of functionality.
- *Adapter agent.* This type of agent is in the central organization of the sensor control. Its function is to try to correct the differences between the measurements of each of the associated sensors to the sensor agents. Thus, a joint database among all recognition systems participating in the architecture is achieved, expanding the source of knowledge.

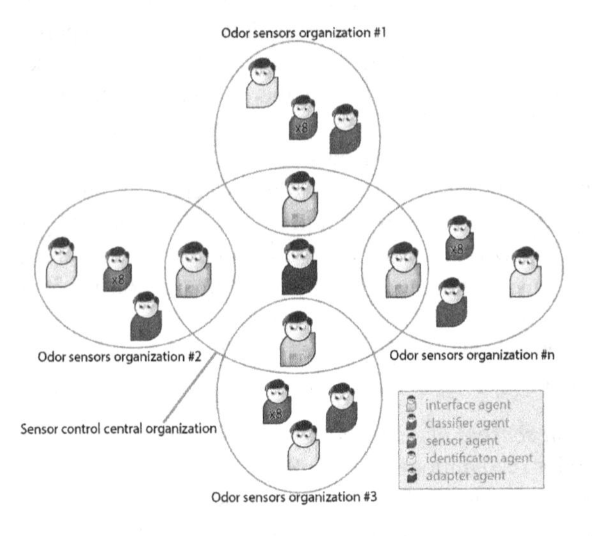

Fig. 1. Structure of virtual organizations of the case study in PANGEA

2.2 Algorithms: Classification Method of Odor Data (e-BPNN)

We will show two types of neural networks: one is a multi-layered neural network based on error back-propagation method and the other is a learning vector quantization (LVQ).

First, we will explain the multi-layered neural networks. In order to classify the odors, we adopt a three-layered neural network based on the error back-propagation method, as shown in Fig.2. The error back-propagation algorithm, which is based on the gradient method, is given by the following steps.

- *Step* 1. Set the initial values of $w_{ji}, w_{kj}, \theta_j, \theta_k$, and $\eta(>0)$.
- *Step* 2. Specify the desired values of the output $d_k, k = 1,2,\cdots,K$ corresponding to the input data x_i, i= 1,2, . . . , I in the input layer.
- *Step* 3. Calculate the outputs of the neurons in the hidden layer by

$$net_j = \sum_{i=1}^{I} w_{ji}x_i - \theta_j \,, O_j = f(net_j), f(x) = \frac{1}{1+e^{-x}}$$

- *Step* 4. Calculate the outputs of the neurons in the output layer by

$$net_k = \sum_{k=1}^{K} w_{kji} O_j - \theta_j \,, O_k = f(net_k), f(x) = \frac{1}{1+e^{-x}}$$

- *Step* 5. Calculate the error *en* and generalized errors by

$$e_k = d_k - O_k$$
$$\delta_k = \delta_k O_k (1 - O_k)$$
$$\delta_j = \sum_{k=1}^{K} \delta_k \, w_{kj} O_j (1 - O_j)$$

- *Step* 6. Use the following formula to calculate half of the sum of the squares of the errors in the output of all.

$$\mathrm{E} = \frac{1}{2} \sum_{k=1}^{K} {e_k}^2$$

- Step 7. If E is sufficiently small, exit the learning. Otherwise, modify the weight by the following equation:

$$\Delta w_{kj} \equiv w_{kj}(t+1) - w_{kj}(t) = \eta \delta_j O_{jk} \qquad \Leftarrow w_{kj} + \Delta w_{kj}$$
$$\Delta w_{ji} \equiv w_{ji}(t+1) - w_{ji}(t) \quad = \eta \delta_i O_j \qquad \Leftarrow w_{ji} + \Delta w_{ji}$$

- Step 8. Go to *Step* 3.

Using the above recursive procedure, we can train the odor data.

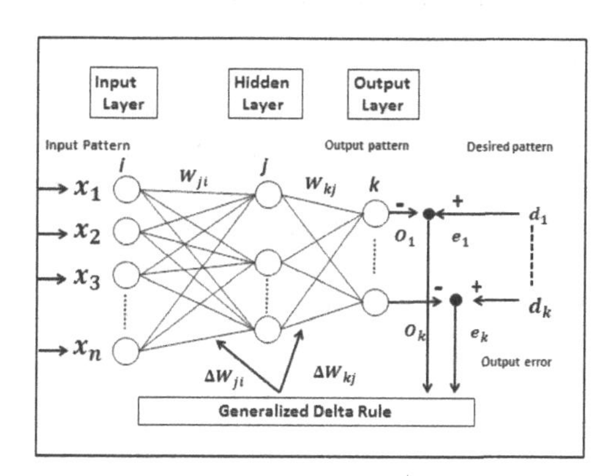

Fig. 2. Three layered neural network with the error back-propagation

The neural network (Fig. 2) consists of three layers: input layer i, hidden layer j and output layer k. When the input data $x_i, i = 1,2,\ldots,I$, are applied in the input layer, we can obtain the output O_k in the output layer, which is compared to the desired value d_k which has been assigned in advance. If the error $e_k = d_k - O_k$

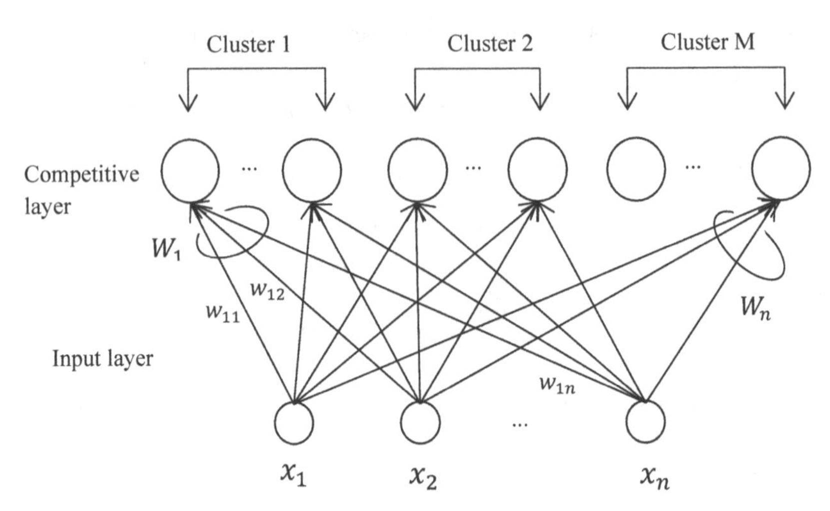

Fig. 3. LVQ structures

occurs, then the weighting coefficients w_{ji}, w_{kj} are corrected so that the error becomes smaller based on the error back-propagation algorithm.

Next, we will show the LVQ:

The structure of LVQ is two layered, consisting of an input layer and a competitive layer as shown in Fig. 3.

In order to classify the odors we adopt a two-layered neural network based on the learning vector quantization method as shown in Fig. 3. Learning vector quantization is a supervised learning method for the purpose of pattern classification for input data. The learning method is given by the following steps:

Step 1. Set the initial values of w_{ij} $(j=1,2,..M, i=1,2,..,n)$, T, and α_0 (> 0) where T is the total iteration number for learning, n is the number of input, M is the number of neurons in cluster j, and α_0 is the initial value of the learning rate.

Step 2. First, calculate proximity to the coupling coefficient vector WJ of the input vector x and neuron j in the sense of Euclidean distance. The neuron with the closest coupling coefficient in the sense of Euclidean distance in the competitive layer neuron is detected by following equation for the input pattern.

$$d_j = \| \mathrm{x} - w_j \| = \sqrt{\sum_{i=1}^{n} (x_i - w_{ji})^2}$$

$$d_c = \| \mathrm{x} - w_c \| = \min_j d_j$$

Step 3. If the input vector and winning neuron c belong to the same class, then change $w_j(t)$ by using the following equation:

$$w_j(t+1) = w_j(t) + \alpha(t)\left(x - w_j(t)\right), j = c$$
$$w_j(t+1) = w_j, j \neq c$$

where

$$\alpha(t) = \alpha_0 \left(1 - \frac{1}{T}\right).$$

If the input vector and neuron c belong to the different class, then change $w_j(t)$ by using the following equation:

$$w_j(t+1) = w_j(t) - \alpha(t)\left(x - w_j(t)\right), j = c$$

$$w_j(t+1) = w_j, j \neq c.$$

Step 4. If t<T, Go to *Step 2*

Using the above recursive procedure, we can train the odor data.

2.3 Principle of QCM Sensors

The QCM has been well-known to provide very sensitive mass-measuring devices in nanogram levels. Synthetic polymer-coated QCMs have been studied as sensors for various gas works as a chemical sensor. The QCM sensors are made by covering the surface with several kinds of a very thin membrane with about 1 mm, as shown in Fig. 4. The QCM sensor is integrated into a resonance circuit. If the film absorbs the odor molecules, the oscillation frequency is reduced since the mass of the vibrator is changed.

Therefore, the frequency (of the QCM) will change according to the deviation of the weight due to the adsorbed odor molecular (odorant). In this paper we have used the materials shown in Table 1. The basic approach used here is a sol-gel method. The process is a wet-chemical technique used for the fabrication of both glassy and ceramic materials. The manufacturing of a film was done by the following procedure.

MTMS (1): Trimethylsilane, ethanol, water and nitric acid. (2) Once stirred, 2-ethyl acrylate (PFOEA) was added to a solution of 1.

Here, HTMS: $CH_3S_i(OCH_3)_3$ and PFOEA: $F(CF_2)_8CH_2CH_2 - OCOCH = CH_2$

Table 1. Chemical materials used as the membrane.We used seven sensors using a solution of three types of in this paper.

$\alpha = C_4H_{12}O_3Si(0.5g), C_{16}H_{19}F_{17}O_3Si(0.015g), 30\%HNO_3(10\mu L), H_2O(0.30ml), etanol(3.0ml)$
$\beta = C_4H_{12}O_3Si(0.5g), C_{16}H_{19}F_{17}O_3Si(0.030g), 30\%HNO_3(10\mu L), H_2O(0.15ml), etanol(3.0ml)$
$\gamma = C_4H_{12}O_3Si(0.5g), 30\%HNO_3(10\mu L), H_2O(0.3ml), etanol(3.2ml)$

Sensor number	Materials of mebrane
sensor1	α
sensor2	β
sensor3	γ
sensor4	α
sensor5	γ
sensor6	α
sensor7	β

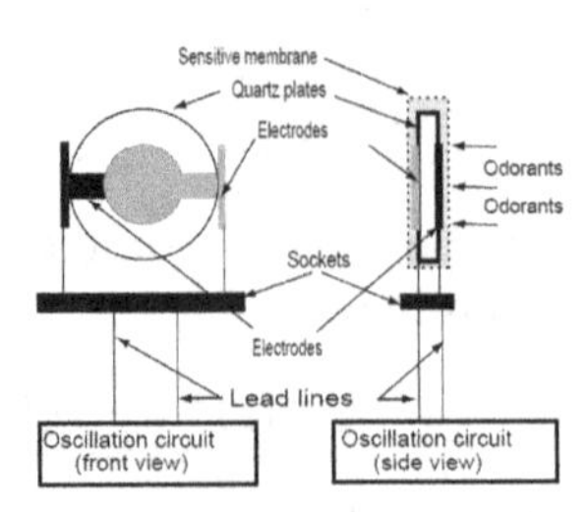

Fig. 4. Principle of QCM sensor

The QCM sensor is integrated into a resonance circuit. If the membrane of the sensor absorbs odorant, the weight of quartz plate will be modified. Oscillation frequency is also reduced when the membrane absorbs odorants. Thus, the original frequency of the crystal oscillation will become smaller according to the density of odorants

2.4 Odor Sensing Hardware System

Generally, this type of system, for example an air purifier, a breathalyzer, etc., is designed to detect some specific odor. Each of the QCM membranes has its own characteristics in response to different odors. When combining many QCM sensors together, the ability to detect the odor is increased. Odor sensing systems (called electronic nose (EN) systems), shown in Fig. 5, were developed based on the concept of the human olfactory system. Odor generating machines are using Permeater, as shown in Fig. 6.

Permeater can continuously generate standard gas of many kinds (inorganic and organic gas) for a long time. The process takes place in the standard gas generator, which generates a continuous standard of trace gas concentration. The combination of QCM sensors, listed in Table 1, is used as the olfactory receptors in the human nose. The odors used here are shown in Table 2.

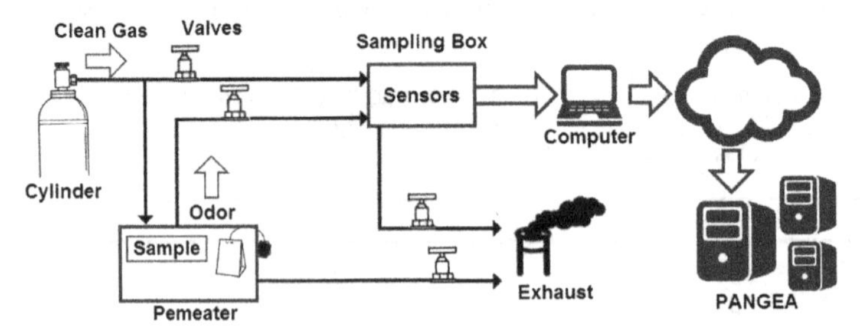

Fig. 5. Odor sensing systems

First, allow the dry air to flow so the odor coming from the Permeater get to the sampling box. Then, the reaction of the QCM sensor is accumulated into frequency counter as data. In addition, the gas is exhausted from the sampling box. The PC converts the data stored in the frequency counter into a text file.

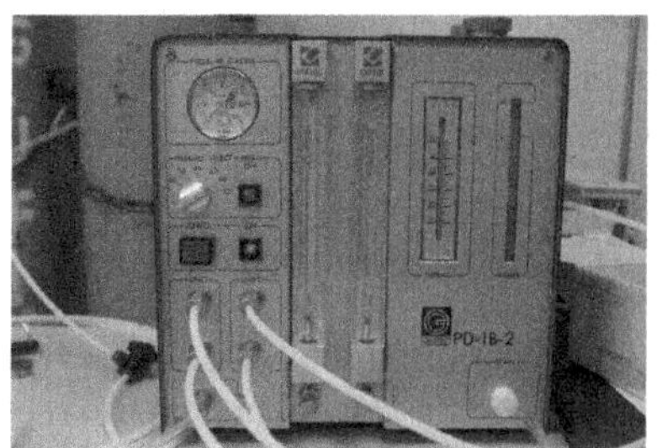

Fig. 6. Permeater and QCM sensors

Table 2. Kinds of odors measured in this experimet

Symbols	Kind of odors
A	ethanol
B	toluen
C	ethly acetatate

2.5 Measurement of Odor Data

We have measured four types of odors as shown in Table 2. The sampling frequencies are 1[Hz]. Diffusion tubes are used to control the density of gases. This is because it is possible to generate the gas at various concentrations by using a diffusion tube through Permeater. Odor data are measured for 900 [s]. They may include impulsive noises due to the typical phenomena of QCM sensors. To remove these impulsive noises we adopt a median filter which replaces a value at a specific time by a median value among neighboring data around the specific time. In Fig. 7 we show the measurement data for the symbol A where the horizontal axis is the measurement time and the vertical axis is the frequency deviation from the standard value (20M[Hz]) after passing through a three-point median filter.

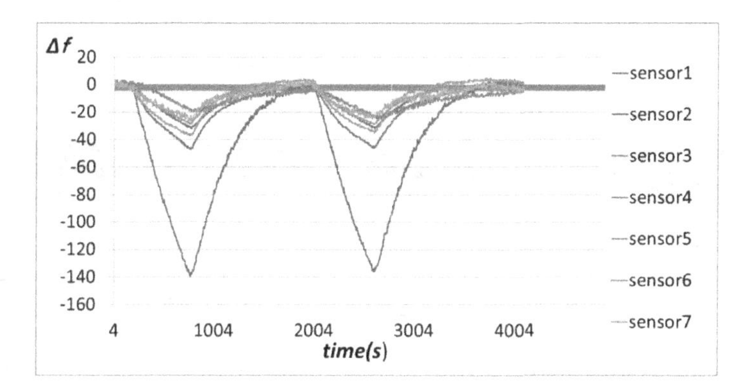

Fig. 7. Measurement of odor data

Here, seven sensors are used. The maximum value for each sensor among seven sensors is selected as a feature value for the sensor. Therefore, for one odor, there are seven sensor values, which will be used for classification.

3 Conclusions and Results

In order to classify the feature vector by using error-back propagation, we allocate the desired output for the input feature vector, which is a seven-dimensional vector, as shown in Table 3. By adding the coefficient of variation to the usual feature vector, the variations for odors are reduced. The training was performed until the total error was less than or equal to 1×10–2 where η=0.8.

Table 3. Training data set for ethanol (A), toluene (B), and ethyl acetate (C)

Symbols	Output A	Output B	Output C
A	1	0	0
B	0	1	0
C	0	0	1

We have examined two algorithms, a learning vector quantization, and error back propagation. In learning vector quantization and error back propagation, the training sample number P'= 8 and test sample number is three.

The total number of classification of 100 test samples is checked. The results are summarized in Table 4 and Table 5.

Table 4. Classification results for learning vector classification (LVQ)

Odor data	Classification results (97%)			
	A	B	C	Correct
A	100	0	0	100
B	1	97	2	97
C	4	1	95	95

Table 5. Classification results for layered neural networks

Odor data	Classification results (88%)			
	A	B	C	Correct
A	86	14	0	86
B	14	84	2	84
C	2	3	95	95

We have presented the reliability of a new EN system designed from various kinds of QCM sensors. We have shown that after training the neural network for each odor, we were able to classify the original odor from the mixed odors in the case of two odors.

In addition, we have proposed and developed a multi-agent system which is able to increment the percentage of correct outputs and reduce the training time by sharing all the data between other similar odor detecting systems, and correcting the error between their sensors. The multi-agent system implemented in PANGEA performs communication, control and data management services in a distributed and flexible way. In the case study presented, a classification method is implemented. However the use of PANGEA makes it possible to extend the system by using new methods for the classification of odors and making the system scalable in terms of functionality.

Acknowledgements. This work has been carried out by the project *Sociedades Humano-Agente: Inmersión, Adaptación y Simulación*. TIN2012-36586-C03-03. Ministerio de Economía y Competitividad (Spain). Project co-financed with FEDER funds.

References

[1] Want, R., Hopper, A., Falcao, V., Gibbons, J.: The Active Badge Location System. ACM Transactions on Information Systems 10(1), 91–102 (1992)
[2] Lim, C.H., Anthony, P., Fan, L.C.: Applying multi-agent system in a context aware. Borneo Science 24, 53–64 (2009)
[3] Ahmed, A., Ali, J., Raza, A., Abbas, G.: Wired Vs Wireless Deployment Support For Wire-less Sensor Networks. In: 2006 IEEE Region 10 Conference TENCON 2006, pp. 1–3 (2006)
[4] Lee, M., Yoe, H.: Comparative Analysis and Design of Wired and Wireless Integrated Networks for Wireless Sensor Networks. In: Software Engineering Research, Management & Applications, SERA 2007, pp. 518–522 (2007)
[5] Bajo, J., Corchado, J.M., De Paz, Y., De Paz, J.F., Rodríguez, S., Martin, Q., Abraham, A.: SHOMAS: Intelligent Guidance and Suggestions in Shopping Centres. Applied Soft Com-puting. Elsevier Science 9(2), 851–862 (2009)
[6] Fonseca, S.P., Martin, P., Griss, L., Letsinger, R.: An Agent Mediated E-Commerce Environment for the Mobile Shopper, Hewlett-Packard Laboratories, Technical Report, HPL-2001-157 (2001)
[7] Kowalczyk, R., Ulieru, M., Unland, R.: Integrating Mobile and Intelligent Agents in Advanced E-commerce: A Survey. In: Kowalczyk, R., Müller, J.P., Tianfield, H., Unland, R. (eds.) NODe-WS 2002. LNCS (LNAI), vol. 2592, pp. 295–313. Springer, Heidelberg (2003)
[8] Carabelea, C., Boissier, O.: Multi-Agent Platforms on Smart Devices: Dream or Reality? In: Proceedings of the SOC 2003, pp. 126–129 (2003)
[9] Fok, C., Roman, G., Lu, C.: Agilla, A Mobile Agent Middleware for Self-Adaptive Wireless Sensor Networks. ACM Transactions on Autonomous and Adaptive Sys. (3), 1–26 (2009)
[10] Meyer, G.G., Främling, K., Holmström, J.: Intelligent Products: A survey. Computers in Industry 60(3), 137–148 (2009)
[11] Kazanavicius, E., Kazanavicius, V., Ostaseviciute, L.: Agent-based framework for embedded systems development in smart environments. In: Proceedings of the 15th International Conference on Information and Software Technologies IT, pp. 194–200 (2009)
[12] Tynam, R., Ruzzelli, A., O'Hare, G.M.P.: A Methodology for the Deployment of Multi-Agent Systems on Wireless Sensor Networks. Mutiagent and Grid Systems, 491–503 (2006)
[13] Rogers, A., Jennings, N.R., Corkill, D.: Agent Technologies for Sensor Networks. IEEE Intelligent Systems 24(2), 13–17 (2009)

[14] Griol, D., García-Herrero, J., Molina, J.M.: Combining heterogeneous inputs for the devel-opment of adaptive and multimodal interaction systems. Advances in Distributed Computing And Artificial Intelligence Journal (2013) ISSN: 2255-2863
[15] Tapia, D.I., Abraham, A., Corchado, J.M., Alonso, R.S.: Agents and ambient intelligence: case studies. Journal of Ambient Intelligence and Humanized Computing 1(2), 85–93 (2010)
[16] Bajo, J., Corchado, J.M.: Evaluation and monitoring of the air-sea interaction using a CBR-Agents approach. Case-Based Reasoning Research and Development 50 (2005)
[17] Corchado, J.M., De Paz, J.F., Rodríguez, S., Bajo, J.: Model of experts for decision support in the diagnosis of leukemia patients. Artificial Intelligence in Med. 46(3), 179–200 (2009)
[18] De Paz, J.F., Rodríguez, S., Bajo, J., Corchado, J.M.: Case-based reasoning as a decision support system for cancer diagnosis: A case study. International Journal of Hybrid Intelligent Systems 6(2), 97–110 (2009)
[19] Tapia, D.I., Rodríguez, S., Bajo, J., Corchado, J.M.: FUSION@, a SOA-based multi-agent archi-tecture. In: International Symposium on Distributed Computing and Artificial Intelligence 2008 (DCAI 2008), pp. 99–107 (2008)
[20] Corchado, J.M., Aiken, J.: Hybrid artificial intelligence methods in oceanographic forecast models. IEEE Transactions on Systems, Man, and Cybernetics, Part C: Applications and Reviews 32(4), 307–313 (2002)
[21] Bajo, J., De Paz, J.F., Rodríguez, S., González, A.: Multi-agent system to monitor oceanic envi-ronments. Integrated Computer-Aided Engineering 17(2), 131–144 (2010)
[22] Hardy, A., Bouhafs, F., Merabti, M.: A Survey of Communication and Sensing for Energy Management of Appliances. International Journal of Advanced Engineering Sciences and Technologies 3(2), 61–77 (2010)
[23] Zato, C., Villarrubia, G., Sánchez, A., Bajo, J., Corchado, J.M.: PANGEA: A New Platform for Developing Virtual Organizations of Agents. International Journal of Artificial Intelligence 11(A13), 93–102 (2013)
[24] Omatu, S., Yano, M.: Intelligent Electronic Nose System Independent on Odor Concentration. In: Abraham, A., Corchado, J.M., González, S.R., De Paz Santana, J.F. (eds.) International Symposium on DCAI. AISC 91, pp. 1–9. Springer, Heidelberg (2011)
[25] Ferber, J., Gutknecht, O., Michel, F.: From Agents to Organizations: an Organizational View of Multi-Agent Systems. In: Giorgini, P., Muller, J., Odell, J. (eds.) AOSE 2003. LNCS, vol. 2935, pp. 214–230. Springer, Heidelberg (2004)
[26] Rodríguez, S., de Paz, Y., Bajo, J., Corchado, J.M.: Social-based planning model for multiagent systems. Expert Systems with Applications 38(10), 13005–13023 (2011)
[27] Foster, I., Kesselman, C., Tuecke, S.: The anatomy of the grid: Enabling scalable virtual organizations. Int. J. High Perform. Comput. Appl. 15(3), 200–222 (2001)
[28] Pavon, J., Sansores, C., Gomez-Sanz, J.J.: Modelling and simulation of social systems with INGENIAS. International Journal of Agent-Oriented Software Engineering 2(2), 196–221 (2008)
[29] Garijo, F., Gómes-Sanz, J.J., Pavón, J., Massonet, P.: Multi-agent system organization: An engineering perspective. In: Pre-Proceeding of the 10th European Workshop on Modeling Autonomous Agents in a Multi-Agent World, MAAMAW (2001)
[30] Milke, J.A.: Application of Neural Networks for discriminating Fire Detectors. In:10 th International Conference on Automatic Fire Detection, AUBE 1995, Duisburg, Germany, pp. 213–222 (1995)
[31] Charumporn, B., Yoshioka, M., Fujinaka, T., Omatu, S.: Identify household burning smell using an electronic nose system with artificial neural networks. In: Proceedings of the IEEE International Symposium on Computational Intelligence in Robotics and Automation, vol. 3, pp. 1070–1074. IEEE (2003)
[32] Fujinaka, T., Yoshioka, M., Omatu, S., Kosaka, T.: Intelligent Elec-tronic Nose Systems for Fire Detection Systems Based on Neural Networks. In: The Second International Conference on Advanced Engineering Computing and Applications in Sciences, ADVCOMP 2008, pp. 73–76. IEEE (2008)

Author Index